AF363299

Diversity of Coral-Associated Fauna II

Diversity of Coral-Associated Fauna II

Editor

Simone Montano

MDPI • Basel • Beijing • Wuhan • Barcelona • Belgrade • Manchester • Tokyo • Cluj • Tianjin

Editor
Simone Montano
Department of Earth and
Environmental Sciences
(DISAT), University of
Milano, Bicocca, Italy

Editorial Office
MDPI
St. Alban-Anlage 66
4052 Basel, Switzerland

This is a reprint of articles from the Special Issue published online in the open access journal *Diversity* (ISSN 1424-2818) (available at: https://www.mdpi.com/journal/diversity/special_issues/Cor_Fau).

For citation purposes, cite each article independently as indicated on the article page online and as indicated below:

LastName, A.A.; LastName, B.B.; LastName, C.C. Article Title. *Journal Name* **Year**, *Volume Number*, Page Range.

ISBN 978-3-0365-5543-0 (Hbk)
ISBN 978-3-0365-5544-7 (PDF)

Cover image courtesy of Bert W. Hoeksema.

Contents

About the Editor

Simone Montano

Simone Montano is a researcher at the University of Milano-Bicocca, Department of Earth and Environmental Sciences (DISAT) in Italy, and Vice-Director of the Marine Research and High Education Center, Magoodhoo Island, Maldives. He is a marine biologist, mainly interested in the ecology and biology of the coral reef ecosystem. His current research activities focus on the assessment of coral health and diseases, with particular attention on new and emerging coral symbioses, and about the emerging field of coral reefs restoration and rehabilitation. In this regard, he is deeply involved in developing new advanced tools to mitigate the impacts of coral diseases and to boost the recovery of the coral reefs. All his activities are aimed to understand the dynamics that will drive this ecosystem under a climate change scenario, in order to develop and propose environmental management plans. To date, he has published more than 90 peer-reviewed papers in international scientific journals. He is a PADI diving instructor with an Advanced European Scientific Diver license, with ¿900 scientific dives and a total of ¿1100. His field work experiences include the Caribbean (St. Eustatius, Bonaire, and Curacao), the Indo-Pacific area (Maldives, Mauritius, Yemen, India, and Thailand), and the Red Sea (Egypt, Saudi Arabia).

Editorial

Diversity of Coral-Associated Fauna: An Urgent Call for Research

Simone Montano [1,2]

1   Department of Earth and Environmental Sciences (DISAT), University of Milano-Bicocca, Piazza Della Scienza, 20126 Milan, Italy; simone.montano@unimib.it
2   MaRHE Center (Marine Research and High Education Center), Magoodhoo Island, Faafu Atoll 12030, Maldives

Citation: Montano, S. Diversity of Coral-Associated Fauna: An Urgent Call for Research. *Diversity* **2022**, *14*, 765. https://doi.org/10.3390/d14090765

Received: 7 September 2022
Accepted: 15 September 2022
Published: 16 September 2022

Publisher's Note: MDPI stays neutral with regard to jurisdictional claims in published maps and institutional affiliations.

Tropical coral reefs are considered the "rainforest of the sea" and are among the marine ecosystems with the highest biodiversity [1]. These "rainforests" are typically composed of assemblages of anthozoans, sponges, bryozoans, and ascidians, forming the three-dimensional matrix which provides architectural complexity for a myriad of organisms [2]. A large proportion of this biodiversity is represented by tiny invertebrates, usually known as cryptofauna, that are often overlooked because of their size, a lack of commercial interest, charisma, and/or taxonomic expertise [3–5]. Despite the fact that nearly a thousand invertebrates are known to depend to some extent on corals for habitat, food, shelter, or settlement cues [6], it is still to be elucidated how many species exhibit obligatory or facultative symbiotic relationships with all other coral reef framework-forming taxa, keeping the coral reefs a mystery in terms of their diversity and functioning.

Unfortunately, increasing evidence suggests that most coral reefs will undergo compositional, structural, and functional changes in response to local stressors, such as overfishing, eutrophication, and diseases [7], as well as in response to global stressors such as ocean acidification and climate-induced coral bleaching, impacting the world's coral reefs and the communities that depend upon them [8]. Indeed, in the future different environmental conditions not only will push some species towards and beyond their physiological limits, but will also modify the network of interactions [9]. Since species are highly interconnected, co-extinction events will have largely unknown ecological consequences as cryptic communities, with largely understudied ecological functions, may disappear accordingly [9]. Thus, it is evident that only by understanding the highly multifaceted interactions amongst the different coral reef organisms will we gain insights on how, where, and why coral reefs are changing [10].

In this respect, this Special Issue aims to provide more testimonies of the extreme diversity of coral-associated fauna, as well as to improve, through different perspectives and new methodological approaches, the knowledge of marine invertebrate diversity in general, and that of the coral reef-associated ones in particular. The contributions published in this volume address specifically a variety of topics including (i) the integrative taxonomy and genetic diversity of merulinid corals [11] and of crustaceans associated with pocilloporid corals [12], (ii) the diversity of coral reef fishes in the Western Indian Ocean [13], (iii) the spatial distribution, host range, and prevalence of associations involving sponges [14], alcyonaceans [15], and corals [16], and (iv) the possible negative impacts that some coral-associated invertebrates can have on the health of their hosts [17–21].

Coral reefs are globally recognized as a major ecosystem in need of conservation [22], which might require the inclusion in future studies of the so far largely ignored hidden or cryptic communities. These communities and their ecological functions are, in fact, markedly different than the visible or exposed ones [23]. Monitoring of coral reef benthos alone is, however, not enough to understand the effects of external drivers on the resilience of coral reefs and their sensitivity to community changes.

Nowadays, it seems appropriate to move forward in adopting new strategies, in addition to multidisciplinary taxonomic studies, to study benthic communities, since this cryptobenthic fauna appears to play a relevant role in the biodiversity and conditions of coral communities. Thus, I advocate for the urgent development and use of emerging new technologies to facilitate and increase the repeatability of coral reef monitoring efforts, as well as to unravel the drivers and feedback mechanisms behind benthic community changes [24]. The systematic collection of data on the coral-associated fauna during free-living biodiversity surveys should strongly be encouraged to increase the knowledge about the diversity of coral-associated fauna, to discover its ecological roles, and to depict its functional traits within coral reefs ecosystems. This process may provide baselines of such hidden biodiversity, identify rare symbiotic species, and it could be used to monitor future changes in symbiotic assemblages. Furthermore, given the paucity of information on the natural history of many symbiont taxa, this approach would also potentially allow us to classify certain symbiont species and help raise awareness of their current endangerment status [5]. Indeed, the easiest and most cost-effective way to protect coral-associated invertebrates will usually be by conserving them alongside their hosts. This represents a paradigm shift from preserving single taxa to protecting symbiont assemblages and micro-ecosystems. However, without in-depth knowledge regarding their hosts range, rate of hosts shift, and symbionts' vulnerability, the level of endangerment of the symbionts may not reflect their real risk of extinction. In addition, novel threats (e.g., diseases and plastic pollution) and their interactions need to be taken into account as they might also play a role in benthic community changes [25]. Thus, the assessment of the health condition of coral-associated fauna should be considered a priority since it may help in understanding the resilience capacity of their host. Moreover, more accurate data on host specificity is strongly necessary for the transition to a new era of solution-oriented science with the potential to prolong the survival of coral populations [26]. In the context of the emerging field of restoration ecology, host translocation and active conservation efforts may threaten the survival of coral-associated species, and of their hosts. Thus, explicit actions to restore coral-associated fauna alongside hosts should be considered as a part of coral restoration planning. Hence, coral reef restoration monitoring plans that allow us to assess the effectiveness of the management approach for both hosts and coral-associated fauna are also needed [27].

In conclusion, as the restructuring of tropical coral reef communities towards different and, perhaps, emergent non-hard-coral-dominated communities becomes inevitable in many locations [10], a more complete overview of host-symbiont associations, the degree of specialization and codependence of these symbiotic relationships, as well as the diversity, distribution, and functional roles of coral and non-coral associated invertebrates is paramount to better understand the dynamics, ecological functions, and societal impacts of these communities.

Funding: This research received no external funding.

Institutional Review Board Statement: Not applicable.

Data Availability Statement: Not applicable.

Acknowledgments: I would like to thank all the authors and referees for their remarkable contributions to this Special Issue.

Conflicts of Interest: The author declares no conflict of interest.

References

1. Fisher, R.; O'Leary, R.A.; Low-Choy, S.; Mengersen, K.; Knowlton, N.; Brainard, R.E.; Caley, M.J. Species Richness on Coral Reefs and the Pursuit of Convergent Global Estimates. *Curr. Biol.* **2015**, *25*, 500–505. [CrossRef] [PubMed]
2. Rossi, S.; Bramanti, L.; Gori, A.; Orejas, C. An overview of the animal forests of the world. In *Marine Animal Forests: The Ecology of Benthic Biodiversity Hotspots*; Springer International Publishing: Cham, Switzerland, 2017; pp. 1–28.
3. Hoeksema, B.W. The hidden biodiversity of tropical coral reefs. *Biodiversity* **2017**, *18*, 8–12. [CrossRef]

4. Montano, S. The extraordinary importance of coral-associated fauna. *Diversity* **2020**, *12*, 357. [CrossRef]
5. Bravo, H.; Xu, T.; van der Meij, S.E. Conservation of Coral-Associated Fauna. In *Earth Systems and Environmental Sciences*; Elsevier: Amsterdam, The Netherlands, 2021; p. 8.
6. Stella, J.S.; Pratchett, M.S.; Hutchings, P.A.; Jones, G.P. Coral-associated invertebrates: Diversity, ecological importance and vulnerability to disturbance. *Oceanogr. Mar. Biol. Annu. Rev.* **2011**, *49*, 43–116.
7. Hughes, T.P.; Kerry, J.T.; Álvarez-Noriega, M.; Álvarez-Romero, J.G.; Anderson, K.D.; Baird, A.H.; Babcock, R.C.; Beger, M.; Bellwood, D.R.; Berkelmans, R.; et al. Global warming and recurrent mass bleaching of corals. *Nature* **2017**, *543*, 373–377. [CrossRef]
8. Hughes, T.P.; Kerry, J.T.; Baird, A.H.; Connolly, S.R.; Dietzel, A.; Eakin, C.M.; Heron, S.F.; Hoey, A.S.; Hoogenboom, M.O.; Liu, G.; et al. Global warming transforms coral reef assemblages. *Nature* **2018**, *556*, 492–496. [CrossRef]
9. Strona, G. Modelling co-extinction. In *Hidden Pathways to Extinction*; Springer International Publishing: Cham, Switzerland, 2022; pp. 75–99.
10. Reverter, M.; Helber, S.B.; Rohde, S.; de Goeij, J.M.; Schupp, P.J. Coral reef benthic community changes in the Anthropocene: Biogeographic heterogeneity, overlooked configurations, and methodology. *Glob. Chang. Biol.* **2022**, *28*, 1956–1971. [CrossRef]
11. Chen, C.-J.; Ho, Y.-Y.; Chang, C.-F. Re-Examination of the Phylogenetic Relationship among Merulinidae Subclades in Non-Reefal Coral Communities of Northeastern Taiwan. *Diversity* **2022**, *14*, 144. [CrossRef]
12. Alonso-Domínguez, A.; Ayón-Parente, M.; Hendrickx, M.E.; Ríos-Jara, E.; Vargas-Ponce, O.; Esqueda-González, M.d.C.; Rodríguez-Zaragoza, F.A. Taxonomic Diversity of Decapod and Stomatopod Crustaceans Associated with Pocilloporid Corals in the Central Mexican Pacific. *Diversity* **2022**, *14*, 72. [CrossRef]
13. Samoilys, M.; Alvarez-Filip, L.; Myers, R.; Chabanet, P. Diversity of Coral Reef Fishes in the Western Indian Ocean: Implications for Conservation. *Diversity* **2022**, *14*, 102. [CrossRef]
14. Gobbato, J.; Magrini, A.; García-Hernández, J.E.; Virdis, F.; Galli, P.; Seveso, D.; Montano, S. Spatial Ecology of the Association between Demosponges and Nemalecium lighti at Bonaire, Dutch Caribbean. *Diversity* **2022**, *14*, 607. [CrossRef]
15. Canessa, M.; Bavestrello, G.; Bo, M.; Enrichetti, F.; Trainito, E. Filling a Gap: A Population of *Eunicella verrucosa* (Pallas, 1766) (Anthozoa, Alcyonacea) in the Tavolara-Punta Coda Cavallo Marine Protected Area (NE Sardinia, Italy). *Diversity* **2022**, *14*, 405. [CrossRef]
16. Maggioni, G.; Huang, D.; Maggioni, D.; Jain, S.S.; Quek, R.Z.B.; Poquita-Du, R.C.; Montano, S.; Montalbetti, E.; Seveso, D. The Association of Waminoa with Reef Corals in Singapore and Its Impact on Putative Immune- and Stress-Response Genes. *Diversity* **2022**, *14*, 300. [CrossRef]
17. Jandang, S.; Bulan, D.E.; Chavanich, S.; Viyakarn, V.; Aiemsomboon, K.; Somboonna, N. First Report of Potential Coral Disease in the Coral Hatchery of Thailand. *Diversity* **2022**, *14*, 18. [CrossRef]
18. Montano, S.; Aeby, G.; Galli, P.; Hoeksema, B.W. Feeding Behavior of *Coralliophila* sp. on Corals Affected by Caribbean Ciliate Infection (CCI): A New Possible Vector? *Diversity* **2022**, *14*, 363. [CrossRef]
19. Hoeksema, B.W.; Smith-Moorhouse, A.; Harper, C.E.; van der Schoot, R.J.; Timmerman, R.F.; Spaargaren, R.; Langdon-Down, S.J. Black Mantle Tissue of Endolithic Mussels (*Leiosolenus* spp.) Is Cloaking Borehole Orifices in Caribbean Reef Corals. *Diversity* **2022**, *14*, 401. [CrossRef]
20. Hoeksema, B.W.; Harper, C.E.; Langdon-Down, S.J.; van der Schoot, R.J.; Smith-Moorhouse, A.; Spaargaren, R.; Timmerman, R.F. Host Range of the Coral-Associated Worm Snail *Petaloconchus* sp. (Gastropoda: Vermetidae), a Newly Discovered Cryptogenic Pest Species in the Southern Caribbean. *Diversity* **2022**, *14*, 196. [CrossRef]
21. Hoeksema, B.W.; Timmerman, R.F.; Spaargaren, R.; Smith-Moorhouse, A.; van der Schoot, R.J.; Langdon-Down, S.J.; Harper, C.E. Morphological Modifications and Injuries of Corals Caused by Symbiotic Feather Duster Worms (Sabellidae) in the Caribbean. *Diversity* **2022**, *14*, 332. [CrossRef]
22. Bellwood, D.R.; Pratchett, M.S.; Morrison, T.H.; Gurney, G.G.; Hughes, T.P.; Álvarez-Romero, J.G.; Day, J.C.; Grantham, R.; Grech, A.; Hoey, A.S.; et al. Coral reef conservation in the Anthropocene: Confronting spatial mismatches and prioritizing functions. *Biol. Conserv.* **2019**, *236*, 604–615. [CrossRef]
23. Kornder, N.A.; Cappelletto, J.; Mueller, B.; Zalm, M.J.L.; Martinez, S.J.; Vermeij, M.J.A.; Huisman, J.; de Goeij, J.M. Implications of 2D versus 3D surveys to measure the abundance and composition of benthic coral reef communities. *Coral Reefs* **2021**, *40*, 1137–1153. [CrossRef]
24. Obura, D.O.; Aeby, G.; Amornthammarong, N.; Appeltans, W.; Bax, N.; Bishop, J.; Brainard, R.E.; Chan, S.; Fletcher, P.; Gordon, T.A.C.; et al. Coral Reef Monitoring, Reef Assessment Technologies, and Ecosystem-Based Management. *Front. Mar. Sci.* **2019**, *6*, 580. [CrossRef]
25. Lamb, J.B.; Willis, B.L.; Fiorenza, E.A.; Couch, C.S.; Howard, R.; Rader, D.N.; True, J.D.; Kelly, L.A.; Ahmad, A.; Jompa, J.; et al. Plastic waste associated with disease on coral reefs. *Science* **2018**, *359*, 460–462. [CrossRef] [PubMed]
26. Van Oppen, M.J.; Oliver, J.K.; Putnam, H.M.; Gates, R.D. Building coral reef resilience through assisted evolution. *Proc. Natl. Acad. Sci. USA* **2015**, *112*, 2307–2313. [CrossRef] [PubMed]
27. Montano, S.; Siena, F.; Amir, H. *Coral Reef Restoration Monitoring Manual-Maldives*, 1st ed.; University of Milano Bicocca: Milano, Italy, 2022; pp. 43–45.

diversity

Article

Spatial Ecology of the Association between Demosponges and *Nemalecium lighti* at Bonaire, Dutch Caribbean

Jacopo Gobbato [1,2,*], Andrea Magrini [1], Jaaziel E. García-Hernández [3], Francesca Virdis [4], Paolo Galli [1,2,5], Davide Seveso [1,2] and Simone Montano [1,2]

[1] Department of Earth and Environmental Sciences (DISAT), University of Milan—Bicocca, Piazza Della Scienza, 20126 Milan, Italy; a.magrini@campus.unimib.it (A.M.); paolo.galli@unimib.it (P.G.); davide.seveso@unimib.it (D.S.); simone.montano@unimib.it (S.M.)
[2] MaRHE Center (Marine Research and High Education Center), Magoodhoo Island, Faafu Atoll 12030, Maldives
[3] Marine Genomic Biodiversity Laboratory, Department of Marine Sciences, University of Puerto Rico—Mayagüez, P.O. Box 9000, Mayagüez, PR 00681, USA; jaaziel.garcia@upr.edu
[4] Reef Renewal Foundation, P.O. Box 371 Kralendijk, Bonaire Dutch Caribbean; francesca@reefrenewalbonaire.org
[5] University of Dubai, Dubai P.O. Box 14143, United Arab Emirates
* Correspondence: jacopo.gobbato@unimib.it; Tel.: +39-392-5567889

Citation: Gobbato, J.; Magrini, A.; García-Hernández, J.E.; Virdis, F.; Galli, P.; Seveso, D.; Montano, S. Spatial Ecology of the Association between Demosponges and *Nemalecium lighti* at Bonaire, Dutch Caribbean. *Diversity* 2022, 14, 607. https://doi.org/10.3390/d14080607

Academic Editors: Harilaos Lessios and Michael Wink

Received: 25 June 2022
Accepted: 26 July 2022
Published: 28 July 2022

Abstract: Coral reefs are known to be among the most biodiverse marine ecosystems and one of the richest in terms of associations and species interactions, especially those involving invertebrates such as corals and sponges. Despite that, our knowledge about cryptic fauna and their ecological role remains remarkably scarce. This study aimed to address this gap by defining for the first time the spatial ecology of the association between the epibiont hydrozoan *Nemalecium lighti* and the Porifera community of shallow coral reef systems at Bonaire. In particular, the host range, prevalence, and distribution of the association were examined in relation to different sites, depths, and dimensions of the sponge hosts. We report *Nemalecium lighti* to be in association with 9 out of 16 genera of sponges encountered and 15 out of 16 of the dive sites examined. The prevalence of the hydroid–sponge association in Bonaire reef was 6.55%, with a maximum value of over 30%. This hydrozoan has been found to be a generalist symbiont, displaying a strong preference for sponges of the genus *Aplysina*, with no significant preference in relation to depth. On the contrary, the size of the host appeared to influence the prevalence of association, with large tubular sponges found to be the preferred host. Although further studies are needed to better understand the biological and ecological reason for these results, this study improved our knowledge of Bonaire's coral reef cryptofauna diversity and its interspecific associations.

Keywords: coral reef; cryptofauna; sponges; hydrozoa; *Aplysina*; prevalence; symbiosis

1. Introduction

Coral reefs are recognized as one of the most important marine ecosystems on the planet, since they host the highest biodiversity among marine environments [1]. The complex topography created by the living organisms, such as cnidarians and sponges, provides a three-dimensional structure that supports an incredible diversity of organisms, well suited for species interactions and associations [2]. Unfortunately, this fundamental environment is experiencing severe degradation due to the impacts directly related to climate change and anthropogenic activities [3]. As these ecosystems disappear, scientists find themselves racing against time to increase our knowledge of cryptofauna ecological interactions and their potential role in the survival and resilience of the reef ecosystem [4]. For example, hermatypic corals have evolved crucial microbial symbiotic relationships in order to maintain their health status, improve energy production, cope with environmental changes, complete nutrient recycling, have a defense mechanism for predators, or as a protection from potential pathogen agents and coral feeding organisms [5–8]. Additionally, stony corals have also developed a symbiotic association with several distinct phyla that are involved such as Cnidaria, Porifera, Echinodermata, Annelida, Arthropoda, and Mollusca [9].

Hydrozoans are an example of a group of organisms that has been able to develop a plethora of symbiotic relationships with several marine organisms [10,11], including scleractinians and sponges [12–14]. Currently, there are records for a total of 20 hydrozoan families and 50 genera involved in symbiotic associations with different animals worldwide [15]. Sponges emerged as a suitable host due to the constant water filtration, which results in the continuous presence of nutrients that are available to its symbionts [15,16]. In particular, there are six families of hydrozoans that are generally found in relation to sponges (Cytaeididae, Corynidae, Cladonematidae, Tubulariidae, Sphaerocorynidae, and Campanulariidae) [16]. Worldwide, a total of 26 species of hydrozoans have been identified as epibionts of sponges; however, little information is known about most of these associations [16,17].

Bonaire coral reef systems have recently been recognized as one of the most biodiverse, robust, resilient, and healthy ecosystems in the South Caribbean region [18]. In this context, the island serves as an interesting hotspot to study hydrozoan–sponge associations, since sponges are one of the dominant benthic groups on the reef, second only to corals [19]. Recently, several studies have been conducted identifying novel symbiotic relationships between the reef organisms, such as the zoantharian *Parazoanthus axinellae* epibiotic on the sponge of the genus *Axinella* [16], *Pteroclava krempfi* with alcyonaceans [20], the sponge *Agelas conifera* and the agariciid corals *Agaricia agaricites* and *Helioseris cucullata* [21], the coral-gall crab *Opecarcinus hypostegus* and the agariciid *Agaricia undata* [22], crabs of the genus *Platypodiella* and zoantharians of the genus *Palythoa* with the sponge *Niphates digitalis* [23], sponges, scleractinians, ascidians and zoantharians with polychaetes *Spirobranchus* [24,25], and the *Stylaster–Millepora* association first reported in Bonaire [26]. Nevertheless, coral reef-associated fauna remain strongly understudied, and the total number of species of micro- and macro-invertebrates involved in association with other reef organisms in this region remains largely unknown, despite the potential benefit that these cryptic associations may have on the survival and resilience of the coral reef ecosystems [4,8]. One of these understudied organisms is *Nemalecium lighti* (Hargitt, 1924), a common thecate hydroid species belonging to the Haleciidae family that can be found all year round in all tropical waters, constituting one of the most abundant hydroid species [27,28]. *N. lighti* can be usually found on reef rock substrate, on corals, and on sponge surfaces, where it can better exploit the presence of planktonic particles to feed in the water column [29,30]. Its presence seems to have no influence on the functionality of the feeding strategy of the sponge host, as already demonstrated for other hydrozoans species [16,30], but see [31]. Therefore the impact of these associations on the sponges appears negligible, or even beneficial in some cases, as it may act as protection from predators thanks to the hydrozoan nematocysts [30,32].

In light of this, there are few studies that have examined the spatial ecology of crypto invertebrates associated with sponges [33,34]. Therefore, the goal of this study was to investigate and characterize the association of *Nemalecium lighti* with sponges in the coral reefs of Bonaire Island, with particular attention focused on determining the host range, prevalence, and distribution of this association. The results obtained provide a foundation for additional studies aimed at bridging the gap in our understanding concerning the cryptofauna diversity and its fundamental ecological role in coral reef ecosystems.

2. Materials and Methods

Underwater surveys were conducted between May and August 2021 to investigate the prevalence and distribution of *Nemalecium lighti*–sponge associations (Figure 1) in the reef system around Bonaire Island (12°12′ N, 68°35′ W), an area which is entirely protected since 1979 as part of the Bonaire National Marine Park (BNMP) [18].

Figure 1. Two examples of the association between demosponges and *Nemalecium lighti* in Bonaire reef system: *N. lighti* associated with (**a**) *Scopalina ruetzleri* and (**b**) *Ircinia* sp.

Along the west coast of the island, 16 different sites were chosen randomly based on their SCUBA shore-diving accessibility (Figure 2 and Table 1).

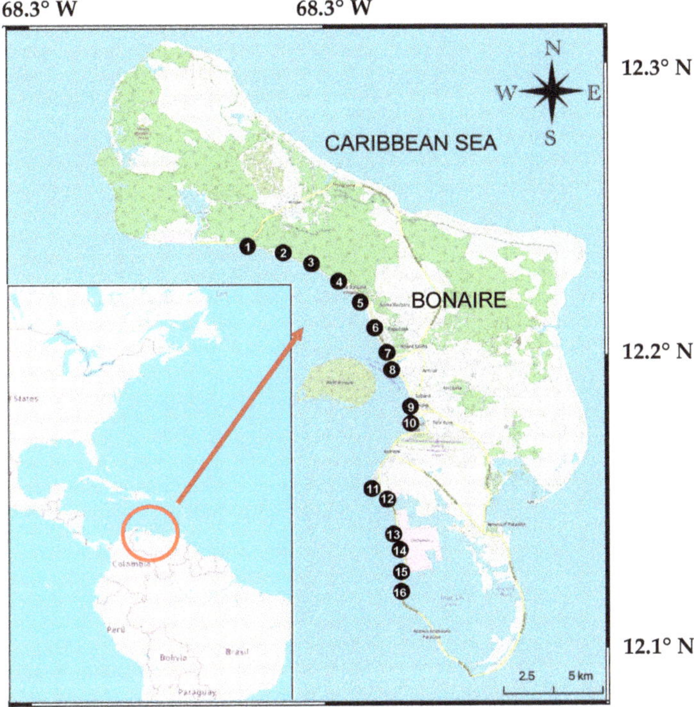

Figure 2. Map of Bonaire, Dutch Caribbean (12°12′ N, 68°35′ W) highlighting the dive sites investigated for sponges–*Nemalecium lighti* association in this study. Map made from OpenStreetMap loaded into QGIS.

Table 1. Coordinates, maximum and mean value of prevalence of association between sponges and *Nemalecium lighti* for each of the dive sites considered for the analyses in the study area.

N°	Dive Sites	Coordinates	Maximum Prevalence (%)	Mean Prevalence (% ± SE)
1	Tolo Reef (Tol)	12°12′92″ N; 068°20′22″ W	15.38	4.89 ± 2.68
2	Jeff Davies Memorial (JDM)	12°12′18″ N; 068°18′50″ W	8.33	2.67 ± 1.69
3	Oil Slick Leap (OSL)	12°12′03″ N; 068°18′51″ W	5.55	0.93 ± 0.93
4	Andrea I (AI)	12°11′29″ N; 068°17′80″ W	12.50	2.08 ± 2.08
5	Andrea II (AII)	12°11′18″ N; 068°17′48″ W	13.33	2.22 ± 2.22
6	La Machaca (LM)	12°10′20″ N; 068°17′22″ W	17.86	8.67 ± 2.73
7	Buddy's Reef (BM)	12°10′14″ N; 068°17′18″ W	16.67	4.01 ± 2.81
8	Bari Reef (BF)	12°10′04″ N; 068°17′10″ W	33.33	20.82 ± 3.56
9	Something Special (SS)	12°09′70″ N; 068°17′02″ W	3.70	2.12 ± 0.70
10	Town Pier (TP)	12°08′57″ N; 068°16′40″ W	2.63	0.44 ± 0.44
11	Punt Viekant (PV)	12°06′91″ N; 068°17′66″ W	0.00	0.00 ± 0.00
12	Alice in Wonderland (AiW)	12°05′99″ N; 068°17′12″ W	22.22	17.59 ± 2.31
13	Salt Pier (SP)	12°05′01″ N; 068°16′91″ W	9.52	4.37 ± 2.01
14	Invisibles (I)	12°04′65″ N; 068°16′80″ W	17.65	2.94 ± 2.94
15	Tori Reef (Tor)	12°04′25″ N; 068°16′84″ W	10.00	1.67 ± 1.67
16	Pink Beach (PB)	12°03′85″ N; 068°16′90″ W	11.11	5.56 ± 1.87

Quantitative analyses were conducted by SCUBA diving, randomly placing three belt transects of 25 m × 2 m at two different depths for each site (total = 96 transects), resulting in 16 "shallow" stations between a 5–9 m depth and 16 "deep" stations between a 10–15 m depth.

Every sponge individual encountered within our transects, including without the presence of *Nemalecium lighti*, was counted. The prevalence was calculated as the number of sponges associated with *N. lighti* divided by the total number of sponges counted at that specific time and place. In addition, the taxon-specific prevalence for each sponge's genus was calculated as the number of sponge hosting associations for each genus, divided by the total number of counted sponges belonging to the same genus, according to Montano et al. 2016 [20]. All sponges were photographed in situ and were identified at the genus level using the relevant literature [35]. Sponges were included in the dataset and counted only when 50% of the individual or more lay within the belt transect area. Furthermore, the potential relationship between the association and the host size was evaluated through a comparison of the observed prevalence with that of five sponge size classes (C1: 5–10 cm; C2: 10–20 cm; C3: 20–30 cm; C4: 40–50 cm; C5: > 50 cm). The size of the sponges was estimated by placing a tape measure on the side of each specimen.

All the data obtained were tested for normality with Kolmogorov–Smirnov tests. In case the normal distribution and homogeneity of variance was violated, Kruskal–Wallis and Mann–Whitney U tests were performed to analyze the mean differences between the sites, depths, and dimensions of the sponge host. Data are presented as the arithmetic mean ± standard error unless stated otherwise. All the statistical analysis performed for this study were conducted using IBM SPSS 27 Software (IBM SPSS 27, New York, NY, USA).

3. Results

In the area investigated, a total of 1755 sponges belonging to the class of Demospongiae were counted and classified in 14 families belonging to 16 different genera, with *Scopalina* (26.67%), *Aplysina* (14.30%), *Agelas* (13.16%), and *Aiolochroia* (9.46%) emerging as the more abundant, while the remaining genera only represented under 7% of the sponges found in the area (Table 2).

Table 2. Genera of sponges considered for the analyses in the study area with values of relative abundance, maximum and mean prevalence of association with *Nemalecium lighti*.

Genus	Relative Abundance (%)	Maximum Prevalence (%)	Mean Prevalence (% ± SE)
Agelas	13.16	7.27	2.70 ± 1.12
Aiolochroia	9.46	8.62	4.09 ± 1.69
Aplysina	14.30	36.17	26.06 ± 4.76
Callyspongia	3.13	0.00	0.00 ± 0.00
Clathria	0.23	0.00	0.00 ± 0.00
Cliona	0.40	25.00	5.00 ± 4.56
Desmapsamma	4.05	8.69	2.74 ± 1.62
Ectyoplasia	3.82	0.00	0.00 ± 0.00
Halisarca	0.17	0.00	0.00 ± 0.00
Iotrochota	4.73	4.00	0.80 ± 0.73
Ircinia	6.38	5.88	2.81 ± 1.01
Monanchora	0.46	0.00	0.00 ± 0.00
Niphates	6.84	0.00	0.00 ± 0.00
Phorbas	0.11	0.00	0.00 ± 0.00
Scopalina	26.67	4.11	1.78 ± 0.76
Verongula	6.09	15.38	6.98 ± 2.34

The genus *Scopalina* resulted as being the most abundant and prevalent group in both the shallow and the deep, with values of, respectively, 27.82% and 26.51% (Table 2). The genus *Agelas* showed a completely different distribution in relation to depth with a relative abundance of 5.45% in the shallow stations and 19.34% in the deep ones (Table 2). A similar trend was observed for *Niphates*, with an increase of the relative abundance from 4.66% in the shallow stations to 8.92% in the deep stations. The genera *Clathria*, *Cliona*, *Halisarca*, *Monanchora*, and *Phorbas* were extremely poorly represented in the study area at both depths, with *Halisarca* resulting as being the only genus completely absent in the shallow stations. Moreover, the spatial distribution of sponges between deep and shallow stations showed different trends for *Agelas*, *Desmapsamma*, *Ectyoplasia*, and *Niphates* at the two depths considered for the study, even if they were not statistically significant (Kruskal–Wallis test, $p > 0.05$).

Regarding the sponge-hydroid interactions, *Nemalecium lighti* has been found in association with 9 out of 16 genera of sponges and in 15 out of 16 of the dive sites surveyed. A total of 115 sponge individuals hosted at least one colony of *N. lighti*. The prevalence of the occurrence of the hydroid–sponge association in the Bonaire reef was 6.55%, with a maximum value of 33.33%. The mean prevalence of the association on the analyzed reef sites was 5.06 ± 1.91%. Among them, "Bari Reef" (BR), "Alice in Wonderland" (AiW), and "La Machaca" (LM) showed higher values with, respectively, 20.82 ± 3.56%, 17.59 ± 2.31%, and 8.67 ± 2.72%, with the others showing less than 6% (Table 1). However, the differences of prevalence among the tested sites were not statistically significant (Kruskal–Wallis test, $p > 0.05$). With regards to the depths considered, the mean prevalence for the deep stations resulted in being higher compared to the shallow stations with, respectively, 5.86 ± 2.77% and 4.27 ± 1.91% (Figure 3a), even if the differences between the stations were not statistically significant both in relation to the site and genera (Mann–Whitney Test, $p > 0.05$).

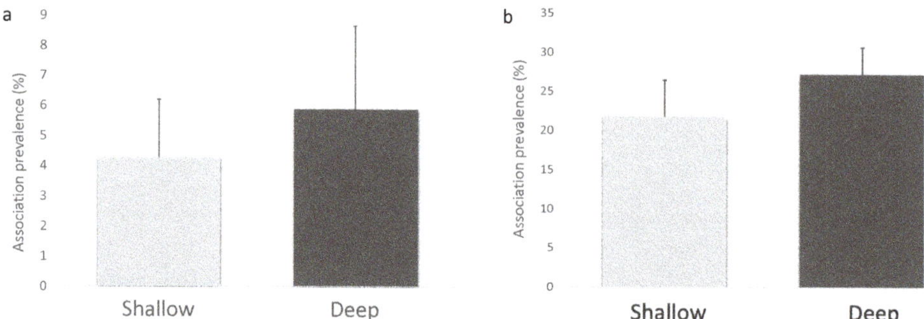

Figure 3. Prevalence of *Nemalecium lighti*–sponges association in the study area: (**a**) Mean association prevalence for all the genera considered and (**b**) for the genus *Aplysina* in relation to depth.

Furthermore, the taxon-specific prevalence was calculated for each genus of sponge that was found to be the host of the association (Table 2). Sponges belonging to the genus *Aplysina* were the most involved in the association with *N. lighti*, with a prevalence of 21.76 ± 4.72% in shallow stations and 27.22 ± 3.44% in the deep stations (Figure 3b). Similarly, *Verongula* and *Aiolochroia* were the second most involved in association with *N. lighti*, even if not notably in the shallow stations. By contrast, *Cliona* showed an elevated prevalence in the shallow station and no association in the deep station. There were seven genera (*Callyspongia*, *Clathria*, *Ectyoplasia*, *Halisarca*, *Monanchora*, *Niphates*, and *Phorbas*) that did not show an association in either the shallow stations or the deep stations, whereas the *Iotrochota*–*Nemalecium lighti* association was found only in deep stations. The differences in the prevalences among the analyzed genera were tested as being statistically significant (Kruskal–Wallis Test, $p < 0.05$).

When correlating the sponge dimension with the *N. lighti* association, most of the sponges belonged to the smaller size classes of 5–30 cm (~70%), whereas only a minor part of the sponges was comprised in the larger size classes of 40–50 cm (Figure 4a). Despite this, a positive increase in prevalence was recorded, with values of 1.41 ± 0.66%, 5.75 ± 0.79%, 8.09 ± 1.62%, 13.78 ± 4.40%, and 15.04 ± 1.96% for C1 to C5 size classes, respectively (Figure 4b). Furthermore, the sponges belonging to the genus *Aplysina* showed a particular behavior in this regard, being the only genus with association cases in all the size classes, in both shallow and deep stations.

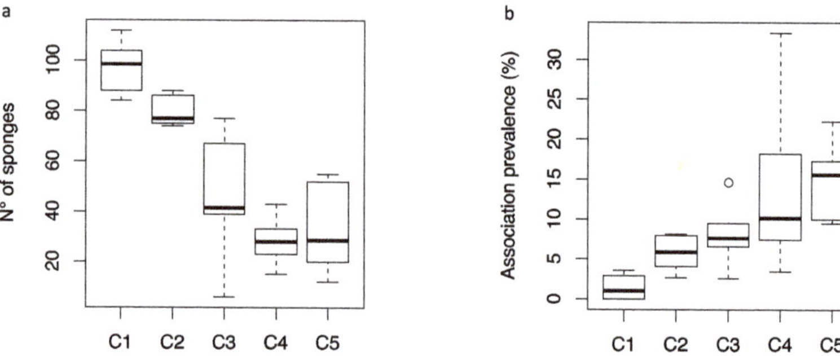

Figure 4. Hosts' size class distribution and prevalence analyses in the study area: (**a**) Number of sponges for each class size; (**b**) Prevalence of *Nemalecium lighti*–sponges association in relation to the host class size. The bold line in the middle of the boxes is the median value, the bottom part of the boxes is the lower quartile, the top part of the boxes is the upper quartile, the lines departing from the boxes are the lower and upper extremes, and the circle is an outlier value.

4. Discussion

Over the years, multiple studies have addressed the role of coral reef biodiversity and its overall impact on the health of these ecosystems [1,4,9]. More specifically, Bonaire coral reef ecosystems emerged as an incredible source of rare or previously unreported reef organism associations. In this perspective, this study assessed for the first time the distribution, host range, and prevalence of the sponges–*Nemalecium lighti* association in this region, aiming to partially fill the knowledge deficiency about cryptofauna associations involving sponges and hydrozoans. The surveys revealed associations between host sponges and *N. lighti* at all the explored sites except one, suggesting a widespread distribution for the association all along the west coast of Bonaire, with some sites showing notably high values of prevalence (BR, AiW, and LM). The host range of this association accounts for nine genera belonging to the class Demospongiae, suggesting that *N. lighti* can be considered as a generalist, since it appears to not target a specific sponge species, at least within the depth ranges we conducted our surveys in. The spatial distribution of the hydroids along the reef zonation has been addressed in previous studies, revealing that different environmental conditions and depths may have an impact on the development of associations and the peculiar assemblage of hydroid species [36–38]. In particular, it has been observed that the maximum diversity of hydrozoans and related associations with sponges was reached on the reef slope, with a continued increase with the depth until reaching 30 m [29]. Similarly, in the present study, the prevalence of associations is slightly higher in the deep stations compared to the shallow ones. However, this result needs to be confirmed in future studies by considering not only the shallower part of the coral reefs but extending the surveys to mesophotic depths. In addition, considering that *N. lighti* is a small cryptic hydroid species growing to a maximum 2 cm that settles on the surface of sponges but sometimes also in protected and shaded parts of their structure, it cannot be excluded that some of the interactions may have been overlooked during our surveys, resulting in an underestimation of its actual prevalence.

The taxon-specific prevalence revealed that the highest prevalence of occurrence of the *N. lighti*–sponge association was observed with the genus *Aplysina* (26%). This prevalence value was almost four times higher than for the second genus involved in the association (*Verongula* with 6.98 ± 2.34%) and extremely more significant than for all the other species, which showed prevalence values lower than 5%. Sponges of the genus *Aplysina* represent an abundant and important component of the marine coastal ecosystems in tropical and subtropical waters [39], where they contribute to the three-dimensionality of the reef structure, which is fundamental for hosting and sustaining several species of associated fauna. The preference of *N. lighti* for this genus may be the result of the complex tubular growth morphology that exposes the epibiont hydroid to a strong current and water flow, characteristics usually exploited by this species [16,40].

In addition, sponges are known to produce an array of chemicals and metabolic products that are ecologically important for different purposes (e.g., growth, protection, competition) [41]. Among them, sponges belonging to the genus *Aplysina* are known to produce high concentrations of brominated alkaloids metabolites (up to 13% of the dry weight) related to antimicrobial activity and cytotoxic activity [42]. These sponges probably produce them as a chemical defense, biofouling, and deterrent against fish predators, as tested on *Thalassoma bifasciatum* and *Blennius sphinx* [43–49]. This peculiar characteristic of the *Aplysina* sponges may be one of the factors that enhances the association with *N. lighti*, as it may take advantage of this defense mechanism of the host to protect itself from predators or microbial offense [50]. However, future investigations to test this hypothesis and to elucidate the nature of this association in the reefs of Bonaire are needed.

Finally, the observed increase in prevalence related to the dimension of the sponges suggests that other factors pertaining to the host also determine the ability of *N. lighti* to settle on sponges. In particular, an increase in sponge (species-dependent) size may correspond to an increase of the favorable surface on which the hydroids can establish the association. In addition, the amount of time necessary for the sponge to grow may also be a contributing factor for the settlement of larvae, increasing the probability of growth on the sponge surface. Further studies are needed to test both of the scenarios proposed and to clearly understand the effect of host size on this particular association.

5. Conclusions

This study provides the first characterization and quantification of the association between the cryptic hydrozoan *Nemalecium lighti* and the demosponge community within the coral reef systems of Bonaire. This hydrozoan has been found to be a generalist symbiont of different genera of sponge along the west coast of Bonaire. Even though the differences in the prevalence of occurrence of the associations were not significantly important with regards to the depth, future investigations extending into the deeper parts of the reef, including mesophotic depths, may be important to better define the role of depth in this association. However, the dimension of the host resulted in influencing the prevalence of association, with a large tubular sponge found to be the preferred host for *N. lighti*. Moreover, the taxon-specific prevalence revealed that the genus most involved in the association with *N. lighti* was *Aplysina*, with an extremely higher prevalence value compared to the other genera recorded. This may be associated with the production of brominated alkaloid metabolites that serve as antimicrobial and biofouling as well as chemical protection from predators. Additional studies are needed to better understand the implications of this preference and what the main biological and ecological reasons for these results are. Overall, this study improves our understanding of the cryptofauna diversity of coral reef associations in Bonaire.

Author Contributions: Conceptualization, J.G. and S.M.; validation, J.E.G.-H., S.M. and D.S.; formal analysis, J.G. and A.M.; investigation, A.M. and F.V.; writing—original draft preparation, J.G.; writing—review and editing, S.M., J.E.G.-H. and D.S.; supervision, S.M., F.V. and P.G. All authors have read and agreed to the published version of the manuscript.

Funding: This research received no external funding.

Data Availability Statement: Not applicable.

Acknowledgments: The authors would like to thank Reef Renewal Bonaire Foundation for the logistic support provided. AM is grateful to Giorgia Ferrari and Camilla Rinaldi for the unique help during the field activities.

Conflicts of Interest: The authors declare no conflict of interest.

References

1. Fisher, R.; O'Leary, R.A.; Low-Choy, S.; Mengersen, K.; Knowlton, N.; Brainard, R.E.; Caley, M.J. Species richness on coral reefs and the pursuit of convergent global estimates. *Curr. Biol.* **2015**, *25*, 500–505. [CrossRef]
2. Stella, J.S.; Jones, G.P.; Pratchett, M.S. Variation in the structure of epifaunal invertebrate assemblages among coral hosts. *Coral Reefs* **2010**, *29*, 957–973. [CrossRef]
3. Caley, M.J.; Buckley, K.A.; Jones, G.P. Separating ecological effects of habitat fragmentation, degradation, and loss on coral commensals. *Ecology* **2001**, *82*, 3435–3448. [CrossRef]
4. Montano, S. The Extraordinary Importance of Coral-Associated Fauna. *Diversity* **2020**, *12*, 357. [CrossRef]
5. Spotte, S. Supply of regenerated nitrogen to sea anemones by their symbiotic shrimp. *J. Exp. Mar. Bio. Ecol.* **1996**, *198*, 27–36. [CrossRef]
6. Rouzé, H.; Lecellier, G.; Mills, S.C.; Planes, S.; Berteaux-Lecellier, V.; Stewart, H. Juvenile *Trapezia* spp. crabs can increase juvenile host coral survival by protection from predation. *Mar. Ecol. Prog. Ser.* **2014**, *515*, 151–159. [CrossRef]
7. Pollock, F.J.; Katz, S.M.; Bourne, D.G.; Willis, B.L. *Cymo melanodactylus* crabs slow progression of white syndrome lesions on corals. *Coral Reefs* **2013**, *32*, 43–48. [CrossRef]
8. Montano, S.; Fattorini, S.; Parravicini, V.; Berumen, M.L.; Galli, P.; Maggioni, D.; Arrigoni, R.; Seveso, D.; Strona, G. Corals hosting symbiotic hydrozoans are less susceptible to predation and disease. *Proc. R. Soc. Biol. Sci.* **2017**, *284*, 20172405. [CrossRef]
9. Stella, J.S.; Pratchett, M.S.; Hutchings, P.A.; Jones, G.P. Coral-associated invertebrates: Diversity, ecological importance, and vulnerability to disturbance. *Oceanogr. Mar. Biol. Annu. Rev.* **2011**, *49*, 43–104.
10. Bo, M.; Di Camillo, C.G.; Puce, S.; Canese, S.; Giusti, M.; Angiolillo, M.; Bavestrello, G. A tubulariid hydroid associated with anthozoan corals in the Mediterranean Sea. *Ital. J. Zool.* **2011**, *78*, 487–496. [CrossRef]

11. Reimer, J.D.; Wee, H.B.; García-Hernández, J.E.; Hoeksema, B.W. Zoantharia (Anthozoa: Hexacorallia) abundance and associations with Porifera and Hydrozoa across a depth gradient on the west coast of Curaçao. *Syst. Biodivers.* **2018**, *16*, 820–830. [CrossRef]

12. Montano, S.; Arrigoni, R.; Pica, D.; Maggioni, D.; Puce, S. New insights into the symbiosis between *Zanclea* (cnidaria, hydrozoa) and scleractinians. *Zool. Scr.* **2015**, *44*, 92–105. [CrossRef]

13. Montano, S.; Seveso, D.; Galli, P.; Puce, S.; Hoeksema, B.W. Mushroom corals as newly recorded hosts of the hydrozoan symbiont *Zanclea* sp. *Mar. Biol. Res.* **2015**, *11*, 773–779. [CrossRef]

14. Montano, S.; Maggioni, D.; Arrigoni, R.; Seveso, D.; Puce, S.; Galli, P. The hidden diversity of *Zanclea* associated with scleractinians revealed by molecular data. *PLoS ONE* **2015**, *10*, e0133084. [CrossRef] [PubMed]

15. Puce, S.; Cerrano, C.; Di Camillo, C.G.; Bavestrello, G. Hydroidomedusae (Cnidaria: Hydrozoa) symbiotic radiation. *J. Mar. Biol. Assoc. UK* **2008**, *88*, 1715–1721. [CrossRef]

16. Puce, S.; Calcinai, B.; Bavestrello, G.; Cerrano, C.; Gravili, C.; Boero, F. Hydrozoa (Cnidaria) symbiotic with Porifera: A review. *Mar. Ecol.* **2005**, *26*, 73–81. [CrossRef]

17. Villamizar, E.; Laughlin, R.A. Fauna Associated with the Sponges *Aplysina archeri* and *Aplysina lacunosa* in a Coral Reef of the Archipelago de Los Roques, National Park, Venezuela. In *Fossil and Recent Sponges*, 1st ed.; Reitner, J., Keupp, H., Eds.; Springer: Berlin/Heidelberg, Germany, 1991; pp. 522–542. [CrossRef]

18. IUCN. Coral Reef Resilience Assessment of the Bonaire National Marine Park. *Netherlands Antilles.* 2011. Available online: https://data.iucn.org/dbtw-wpd/edocs/2011-008.pdf (accessed on 23 May 2022).

19. Frade, P.R.; Bongaerts, P.; Baldwin, C.C.; Trembanis, A.C.; Bak, R.P.M.; Vermeij, M.J.A. Bonaire and Curaçao. In *Mesophotic Coral Ecosystems*, 1st ed.; Loya, Y., Puglise, K., Bridge, T., Eds.; Series Coral Reef of the World; Springer: Cham, Switzerland, 2019; pp. 149–162. [CrossRef]

20. Montano, S.; Allevi, V.; Seveso, D.; Maggioni, D.; Galli, P. Habitat preferences of the *Pteroclava krempfi*-alcyonaceans symbiosis: Inner vs outer coral reefs. *Symbiosis* **2017**, *72*, 225–231. [CrossRef]

21. García-Hernández, J.E.; van Moorsel, G.W.N.M.; Hoeksema, B.W. Lettuce corals overgrowing tube sponges at St. Eustatius, Dutch Caribbean. *Mar. Biodivers.* **2017**, *47*, 55–56. [CrossRef]

22. García-Hernández, J.E.; de Gier, W.; van Moorsel, G.W.N.M.; Hoeksema, B.W. The scleractinian *Agaricia undata* as a new host for the coral-gall crab *Opecarcinus hypostegus* at Bonaire, southern Caribbean. *Symbiosis* **2020**, *81*, 303–311. [CrossRef]

23. Garcia-Hernandez, J.E.; Reimer, J.D.; Hoeksema, B.W. Sponges hosting the zoantharia-associated crab *Platypodiella spectabilis* at St. Eustatius, Dutch Caribbean. *Coral Reefs* **2016**, *35*, 209–210. [CrossRef]

24. García-Hernández, J.E.; Hoeksema, B.W. Sponges as secondary hosts for Christmas tree worms at Curaçao. *Coral Reefs* **2017**, *36*, 1243–1244. [CrossRef]

25. Hoeksema, B.W.; García-Hernández, J.E.; van Moorsel, G.W.N.M.; Olthof, G.; ten Hove, H.A. Extension of the recorded host range of Caribbean Christmas tree worms (*Spirobranchus* spp.) with two scleractinians, a zoantharian, and an ascidian. *Diversity* **2020**, *12*, 115. [CrossRef]

26. Montano, S.; Reimer, J.D.; Ivanenko, V.N.; García-Hernández, J.E.; van Moorsel, G.W.N.M.; Galli, P.; Hoeksema, B.W. Widespread occurrence of a rarely known association between the hydrocorals *Stylaster roseus* and *Millepora alcicornis* at Bonaire, Southern Caribbean. *Diversity* **2020**, *12*, 218. [CrossRef]

27. Gravier-Bonnet, N.; Bourmaud, C.A.F. Hydroids (Cnidaria, Hydrozoa) of Coral Reefs: Preliminary Results on Community Structure, Species Distribution and Reproductive Biology in Juan de Nova Island (Southwest Indian Ocean). *West. Indian Ocean J. Mar. Sci.* **2006**, *5*, 123–132. [CrossRef]

28. Coma, R.; Ribes, M.; Orejas, C.; Gili, J.M. Prey capture by a benthic coral reef hydrozoan. *Coral Reefs* **1999**, *18*, 141–145. [CrossRef]

29. Di Camillo, C.G.; Bavestrello, G.; Valisano, L.; Puce, S. Spatial and temporal distribution in a tropical hydroid assemblage. *J. Mar. Biol. Assoc. UK* **2008**, *88*, 1589–1599. [CrossRef]

30. Bouillon, J.; Gravili, C.; Pagès, F.; Gili, J.M.; Boero, F. *An Introduction to Hydrozoa*, 1st ed.; Betsch, J.M., Bouchet, P., Érard, C., Eds.; Mémoires du Muséum national d'histoire naturelle: Paris, France, 2006.

31. Lewis, T.B.; Finelli, C.M. Epizoic zoanthids reduce pumping in two Caribbean vase sponges. *Coral Reefs* **2015**, *34*, 291–300. [CrossRef]

32. Puyana, M.; Fenical, W.; Pawlik, J.R. Are there activated chemical defenses in sponges of the genus *Aplysina* from the Caribbean? *Mar. Ecol. Prog. Ser.* **2003**, *246*, 127–135. [CrossRef]

33. Wulff, J.L. Ecological interactions of marine sponges. *Can. J. Zool.* **2006**, *84*, 146–166. [CrossRef]

34. Padua, A.; Lanna, E.; Klautau, M. Macrofauna inhabiting the sponge *Paraleucilla magna* (Porifera: Calcarea) in Rio de Janeiro, Brazil. *J. Mar. Biol. Assoc. UK* **2013**, *93*, 889–898. [CrossRef]

35. Zea, S.; Henkel, T.P.; Pawlik, J.R. The Sponge Guide: A Picture Guide to Caribbean Sponges. Available online: https://www.spongeguide.org (accessed on 21 January 2022).

36. Calder, D.R. Local distribution and biogeography of the hydroids (Cnidaria) of Bermuda. *Caribb. J. Sci.* **1993**, *29*, 61–74.

37. Calder, D.R. Associations between hydroid species assemblages and substrate types in the mangal at Twin Cays, Belize. *Can. J. Zool.* **1991**, *69*, 2067–2074. [CrossRef]

38. Calder, D.R. Abundance and distribution of hydroids in a mangrove ecosystem at Twin Cays, Belize, Central America. *Hydrobiologia* **1991**, *216*, 221–228. [CrossRef]

39. Diaz, M.C.; Rützler, K. Sponges: An essential component of Caribbean coral reefs. *Bull. Mar. Sci.* **2001**, *69*, 535–546.

40. Hargitt, C.M. Hydroids of the Philippine Islands. *Philipp. J. Sci.* **1924**, *24*, 467–507. Available online: https://www.marinespecies.org/aphia.php?p=sourcedetails&id=24462 (accessed on 14 July 2022).
41. Amelia, M.; Suaberon, F.A.; Vad, J.; Fahmi, A.D.M.; Saludes, J.P.; Bhubalan, K. Recent Advances of Marine Sponge—Associated Microorganisms as a Source of Commercially Viable Natural Products. *Mar. Biotechnol.* **2022**, *24*, 492–512. [CrossRef]
42. Hentschel, U.; Schmid, M.; Wagner, M.; Fieseler, L.; Gernert, C. Isolation and phylogenetic analysis of bacteria with antimicrobial activities from the Mediterranean sponges *Aplysina aerophoba* and *Aplysina cavernicola*. *FEMS Microbiol. Ecol.* **2001**, *35*, 305–312. [CrossRef]
43. Thoms, C.; Ebel, R.; Proksch, P. Activated chemical defense in *Aplysina* sponges revisited. *J. Chem. Ecol.* **2006**, *32*, 97–123. [CrossRef]
44. Assmann, M.; Lichte, E.; Pawlik, J.R.; Köck, M. Chemical defenses of the Caribbean sponges *Agelas wiedenmayeri* and *Agelas conifera*. *Mar. Ecol. Prog. Ser.* **2000**, *207*, 255–262. [CrossRef]
45. Ebel, R.; Brenzinger, M.; Kunze, A.; Gross, H.J.; Proksch, P. Wound activation of protoxins in marine sponge *Aplysina aerophoba*. *Can. Field-Nat.* **1997**, *111*, 1451–1462. [CrossRef]
46. Weiss, B.; Ebel, R.; Elbrächter, M.; Kirchner, M.; Proksch, P. Defense metabolites from the marine sponge *Verongia aerophoba*. *Biochem. Syst. Ecol.* **1996**, *24*, 1–7. [CrossRef]
47. Teeypant, R.; Woerdenbang, H.J.; Gross, H.J.; Proksch, P. Biotransformation of the Brominated Compounds in the Marine Sponge *Verongia aerophoba*: Evidence for an Induced Chemical Defense? *Planta Med.* **1993**, *59*, A641–A642. [CrossRef]
48. Thoms, C.; Wolff, M.; Padmakumar, K.; Ebel, R.; Proksch, P. Chemical Defense of Mediterranean Sponges *Aplysina cavernicola* and *Aplysina aerophoba*. *Z. Fur Naturforsch.* **2004**, *59*, 113–122. [CrossRef] [PubMed]
49. Chanas, B.; Pawlik, J.R.; Lindel, T.; Fenical, W. Chemical defense of the Caribbean sponge *Agelas clathrodes* (Schmidt). *J. Exp. Mar. Bio. Ecol.* **1997**, *208*, 185–196. [CrossRef]
50. Wulff, J.L. Resistance vs recovery: Morphological strategies of coral reef sponges. *Funct. Ecol.* **2006**, *20*, 699–708. [CrossRef]

Article

Filling a Gap: A Population of *Eunicella verrucosa* (Pallas, 1766) (Anthozoa, Alcyonacea) in the Tavolara-Punta Coda Cavallo Marine Protected Area (NE Sardinia, Italy)

Martina Canessa [1,*], **Giorgio Bavestrello** [1], **Marzia Bo** [1], **Francesco Enrichetti** [1] and **Egidio Trainito** [2]

[1] Dipartimento di Scienze della Terra dell'Ambiente e della Vita (DISTAV), Università di Genova, Corso Europa, 26-16132 Genova, Italy; giorgio.bavestrello@unige.it (G.B.); marzia.bo@unige.it (M.B.); francesco.enrichetti@edu.unige.it (F.E.)

[2] Tavolara Punta Coda Cavallo MPA, Via S. Giovanni, 14-07026 Olbia, Italy; et@egidiotrainito.it

* Correspondence: marti.canessa@gmail.com; Tel.: +39-34-2560-1343

Abstract: Among Mediterranean habitat-forming alcyonaceans, the sea fan *Eunicella verrucosa* is known to form dense forests at circalittoral depths, providing seascape complexity and sustaining a rich associated fauna. Its occurrence in the Tavolara–Punta Coda Cavallo Marine Protected Area (NE Sardinia) has never been deeply investigated despite this area being well known from a biocoenotic point of view. This study provides new information on the size of the colonies settled between 35 and 59 m depth on granitic outcrops and represents a contribution to highlighting the hotspot of megabenthic diversity enclosed in the protected area. The presence of 100 colonies was assessed by photographic samplings performed between 2015 and 2020, in a small area characterized by peculiar ecological conditions. The morphometric descriptions and age estimation showed a persistently isolated population probably derived from a stochastic event of settling of larvae presumably coming from the Tuscany Archipelago. A richly associated epibiotic community, composed of 18 species/OTUs, showed how branched bryozoans, particularly *Turbicellepora avicularis*, and the parasitic octocoral *Alcyonium coralloides*, affected the colonies' branches, suggesting a putative anthropogenic impact related to fishing activity. This study indicates that proper protection and management strategies are mandatory for the Marine Protected Area, in order to conserve this unique population and the whole associated benthic assemblage.

Keywords: *gorgonians*; coralligenous assemblages; fishing impact; Mediterranean sea

Citation: Canessa, M.; Bavestrello, G.; Bo, M.; Enrichetti, F.; Trainito, E. Filling a Gap: A Population of *Eunicella verrucosa* (Pallas, 1766) (Anthozoa, Alcyonacea) in the Tavolara-Punta Coda Cavallo Marine Protected Area (NE Sardinia, Italy). *Diversity* **2022**, *14*, 405. https://doi.org/10.3390/d14050405

Academic Editors: Simone Montano and Michael Wink

Received: 3 April 2022
Accepted: 8 May 2022
Published: 20 May 2022

Publisher's Note: MDPI stays neutral with regard to jurisdictional claims in published maps and institutional affiliations.

1. Introduction

The sea fan *Eunicella verrucosa* (Pallas, 1766) is a large, erected gorgonian, profusely branched, white to deep pink in color, with an Atlanto-Mediterranean distribution, ranging from Angola to Ireland, in the East Atlantic Ocean [1,2]. The species settles directly on bedrock, on large boulders and on artificial surfaces in areas with moderate water movement [3]. The depth range for the Mediterranean Sea goes from 26 to 215 m depth with most of the records located below 35 m depth [1,4–9]. The populations of this species provide structural complexity sustaining rich, associated biodiversity and aesthetic value to sublittoral communities [10,11]. The tridimensional morphology of *E. verrucosa* colonies observed in previous studies is complex and variable: the typical architecture is planar, but other growth forms, characterized by a high rate of branch overlapping, result from different environmental conditions, such as spatial constraints, current intensity, feeding ability and predation [12–18].

The growth rate of *E. verrucosa* was recorded as highly variable: demographic studies of different populations suggest values of 0.6–3.5 cm year^{-1} for Mediterranean areas and of 1–4.5 cm year^{-1} for the English Channel [6,19–22]. The relatively slow growth rate of *E. verrucosa* coupled with its sensitivity to abrasion, mechanical disturbance by anchors, fishing gear and fin-stroke damage by scuba divers [23–26], as well as substratum loss [21,27], makes this species particularly sensitive to anthropogenic impacts and environmental stressors.

Thanks to recent Remotely Operated Vehicle (ROV) surveys [8] and SCUBA dives coupled with citizen science reports, the distribution of *E. verrucosa* on the Mediterranean scale has been recently updated [28]. This large amount of data confirms the occurrence of the species mainly along the coast of the western Mediterranean basin with a peak of records in the Ligurian Sea. In particular, a structured community dominated by this species was identified here in dozens of sites, mainly on sub-horizontal rocks characterized by heavy silting between 30 and 215 m (maximum occurrences 60–90 m) [8]. A lower number of records involve the North African coasts, the Sicily Channel, the North Adriatic Sea and the Aegean Sea. In this scenario, the most impressive gap sees the almost complete absence of the species from the central Tyrrhenian Sea, and in particular, from the Sardinian coast [28].

Very few data are available in the literature regarding NE Sardinia. The oldest records date back to 1990 when Bianchi et al. [29] recorded the species at a site in the Tavolara-Punta Coda Cavallo Marine Protected Area (TPCCMPA). Later, in the same area, field campaigns reported some colonies on the hull of the Klearchos wreck at 77 m depth [30], and another specimen settled on a granitic shoal (Tavolara2) at 55 m depth [31]. Recent surveys assessed the occurrence of this species, in association with large and erect sponge assemblages, on granite reliefs in the Tavolara Channel under 40 m depth [32,33].

This study aims to quantify the presence of *E. verrucosa* in the TPCCMPA, improve the knowledge of its Mediterranean distribution, and provide a morphometric description of the population. In addition, a study of the opportunistic fauna living on the colonies, used as an indicator of mechanical abrasion also of anthropic origin [34,35], was conducted to evaluate the impact of anthropogenic activities within the MPA.

2. Materials and Methods

Between 2015 and 2020, 110 scuba dives were carried out on 77 granite outcrops within the Tavolara Channel at depths between 12 m and 59 m (Figure 1). Each outcrop was georeferenced on a Geographic Information System (GIS) and the coordinates of the sites investigated were registered on the MPA web platform. All the sites where *Eunicella verrucosa* was recorded are listed in Table 1.

Table 1. Investigated sites with the occurrence of *Eunicella verrucosa* within the Tavolara–Punta Coda Cavallo Marine Protected Area and number of specimens in each one. Coordinates of the sites can be consulted upon request to the Marine Protected Area repository.

Site	Depth Range (m)	Outcrop Area (m²)	N Colony	N Dives
N1	37–44	428	6	3
N2	39–45	1336	1	1
N24	38–44	1408	1	1
N25	43–54	677	6	2
N27	38–47	1024	14	6
N28	38–47	3343	4	3
N29	42–49	410	2	1
N30	46–50	435	4	1
N32	47–52	371	3	1
N33	50–56	574	2	1
N34	38–45	421	3	1
N37	35–41	736	3	2
N73	36–40	100	1	1
N74	40–45	345	1	1
N75	40–45	595	1	1
N83	48–52	1022	2	1

Table 1. *Cont.*

Site	Depth Range (m)	Outcrop Area (m²)	N Colony	N Dives
N95	41	221	2	1
N99	45–50	1253	1	2
N100	38–45	498	1	1
N101	46–52	820	1	1
N102	47–53	264	1	1
N118	47–52	580	3	2
N119	39–44	286	1	1
N120	40–45	331	1	1
N140	42–48	900	4	1
N148	48–54	471	3	1
N150	43–49	165	4	1
N151	38–46	198	3	3
N159	48–59	272	12	1
N160	44–49	760	3	1
N165	44–48	197	3	1
N171	40–44	141	1	1
N180	40–45	92	1	1
N182	40–45	280	1	1

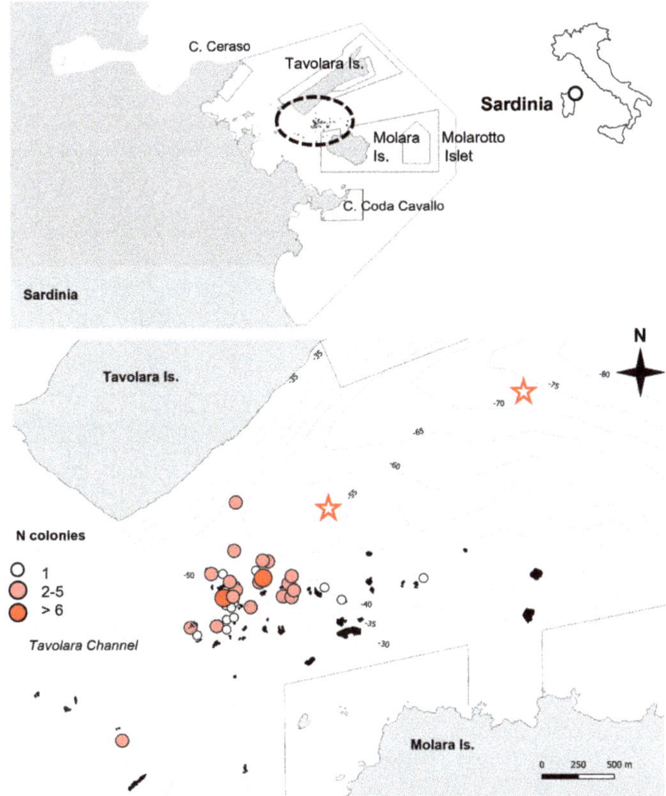

Figure 1. Location of the Tavolara–Punta Coda Cavallo Marine Protected Area and investigated sites (black polygons) with the indication of the number of *Eunicella verrucosa* colonies per site. Stars refer to ANDROMEDE [30,31] reports.

All the recorded colonies of *E. verrucosa* were photographed. The multi-zoom photographic approach [36] was used to characterize the site geomorphologies, the benthic assemblages, and in particular, the presence of *E. verrucosa*, and the occurrence of epibionts and damages on the sea fans (Figure 2).

Figure 2. Operative workflow for the characterization of the investigated sites by a multi-zoom approach. Preliminary Side Scan Sonar survey (**A**) graphic reconstruction and (**B**) final panoramic photographic rendering (**C**) of the site. Examples of the ecological context in which *Eunicella verrucosa* settles, mainly composed of large and erect sponges such as *Axinella* spp., *A. polypoides*, *Spongia lamella* (**D**), *Sarcotragus foetidus* (**E**) and other encrusting and massive species. Details of damages, (**F**) as entangled lines (b) and epibionts, (**G**,**H**) such as *Turbicellepora avicularis*, (c) *Crella elegans* and (e) other acrophilic species associated to the gorgonian (**H**) as *Astrospartus mediterraneus* (d).

Images were taken using a Sony A6000 camera (24 megapixels, two Inon S2000 strobes, color temperature 5000 K) with Sony 16–50 lens (focal length 19 mm), Nauticam WW1 wet wide lens (130° rectilinear field angle) and a Sea & Sea MDX-A6000 underwater case with a flat porthole. Panoramic renderings of the sites to localize the colonies were obtained with multiple shots subsequently joined and optimized in postproduction using the Photoshop CS6 Merge tool.

Using a laser gauge as a reference (wheelbase 25 cm), the height was measured in each colony whereas the fan surface was evaluated for only 63 specimens with a suitable perspective. Photographic processing for measurements was carried out using ImageJ Software (Wayne Rasband and contributors, National Institutes of Health, Bethesda, MD, USA) [37].

As gorgonian colony height is considered a robust parameter leveraged for age estimation in this species [38], this datum was calculated according to the function proposed by Chimienti [39] for Mediterranean populations (Table 2):

$$\text{Age} = e^{\dfrac{H + 18.39}{17.94}} \tag{1}$$

Table 2. Morphometric parameters, percent portion of surface covered by epibionts, naked skeleton and damage of the specimens of *Eunicella verrucosa* investigated in the present study.

Site	Height (cm)	Surface (cm^2)	Age (years)	Epiosis %	Naked Skeleton %	Damage %
N1	41.03	688.07	30.5	5	0	5
N1	39.21	823.51	27.4	15	0	15
N1	43.27	1028.3	34.7	0	0	0
N1	35.02		21.57	0	0	0
N1	19.49	183.14	8.8	0	0	0
N1	36.14		23	0	0	0
N2	27.31	490.01	13.8	30	0	30
N24	10.73	26.92	5.3	0	0	0
N25	26.68		13.4	0	0	0
N25	29.12		15.4	5	0	5
N25	27.4		13.9	0	0	0
N25	46.34		41.4	20	0	20
N25	48.64		47.2	0	100	100
N25	52.3		58.3	40	0	40
N27	20.5	176.32	9.4	0	0	0
N27	31.43	423.55	17.5	0	0	0
N27	24.04		11.5	65	0	65
N27	48.3		46.3	10	0	10
N27	37.27		24.5	0	0	0
N27	34.45	1985.9	20.9	15	0	15
N27	24.93	490.38	12.1	10	0	10
N27	39.55	823.07	28.0	5	0	5
N27	34.5	595.65	20.9	0	0	0
N27	20		9.1	30	5	35
N27	55.64		70.6	15	0	15
N27	16.86		7.6	0	0	0
N27	40.7		29.9	20	0	20
N27	40.7	267.05	29.9	15	0	15
N28	41.53	220.39	31.4	0	0	0
N28	48.32	382.91	46.3	0	0	0
N28	40.65	1219.6	29.8	0	0	0
N28	34.67	811.27	21.1	0	0	0
N29	36.14		23.0	0	0	0
N29	34.46		20.9	5	0	5
N30	66.32	1375.7	130.5	0	0	0
N30	53.49	1688.0	62.4	20	5	25
N30	19.32		8.7	85	0	85
N30	36.13	932.27	23.0	0	0	0
N32	15.11		6.9	50	0	50
N32	55.13	1261.8	68.6	20	5	25
N32	50.98	1085.5	54.0	0	0	0
N33	25.76		12.7	0	0	0
N33	26.31	233.29	13.1	0	0	0
N34	32.34	528.52	18.5	50	40	90
N34	32.04	338.25	18.2	5	0	5
N34	55.65	1892.9	70.6	15	0	15
N37	16.85		7.6	5	0	5
N37	18.22		8.2	0	0	0
N37	14.61		6.7	0	50	50
N73	19.8	299.26	9.0	50	0	50

Site	Height (cm)	Surface (cm^2)	Age (years)	Epiosis %	Naked Skeleton %	Damage %
N74	22.56		10.5	0	0	0
N75	59.06		85.9	0	0	0
N83	40.04	610.01	28.8	0	0	0
N83	18.17	291.49	8.2	0	0	0
N95	48.59	1873.8	47.1	0	0	0
N95	34.2	588.98	20.6	0	0	0
N99	25.37	397.06	12.4	40	0	40
N100	35.7	713.1	22.5	0	0	0
N101	35.72		22.5	0	0	0
N102	27.73	230.85	14.2	0	0	0
N118	51.03	2344.6	54.2	25	0	25
N118	45.04		38.4	0	0	0
N118	12.355		5.9	80	0	80
N119	18.9	159.4	8.5	0	0	0
N120	10			0	0	0
N140	63.55		111.3	0	0	0
N140	55.9	1754.9	71.7	5	0	5
N140	36.22		23.1	0	0	0
N140	48.7	1222.9	47.4	5	0	5
N148	33.31	722.6	19.5	0	0	0
N148	46.43	1302.1	41.6	5	0	5
N148	36.8	821.99	23.9	5	0	5
N150	39.02	492.64	27.1	5	0	5
N150	26.4		13.1	0	0	0
N150	28.2		14.6	20	0	20
N150	19.95		9.1	0	0	0
N151	41.16	1318.0	30.7	0	0	0
N151	49.86	982.98	50.6	10	10	20
N151	36.59	691.24	23.6	0	0	0
N159	29.67	539.54	15.9	5	0	5
N159	43.78	668.55	35.7	0	0	0
N159	38.33	492.06	26.1	0	0	0
N159	37.72	733.66	25.2	0	0	0
N159	18.24	169.71	8.2	0	0	0
N159	32.6	765.03	18.8	0	0	0
N159	34.21	693.07	20.6	0	0	0
N159	46.09	1832.4	40.8	30	0	30
N159	37.02	49.97	24.2	0	0	0
N159	23.59		11.2	0	0	0
N159	37.58	1325.2	25.0	10	0	10
N159	48.11	1326.8	45.8	5	0	5
N160	35.54	720.5	22.2	0	0	0
N160	43.31	495.95	34.7	0	0	0
N160	43.42	863.23	35.0	0	0	0
N165	56.75	1796.2	75.3	5	0	5
N165	40.96	703.3	30.4	0	0	0
N165	35.5		22	0	0	0
N171	43.93		36.0	5	5	10
N180	26.42	385.42	13.2	5	0	5
N182	31.48	599.32	17.6	5	0	5

The size and age structure of the population were analysed in terms of size–frequency and distribution parameters (skewness and kurtosis) using Past 4.10 statistic (Øyvind Hammer, Natural History Museum, University of Oslo, Oslo, Norway).

Finally, photographs were analyzed to identify associated epibiont species to the lowest possible taxonomic level: when the identification was not possible, Operational Taxonomic Units (OTUs) were adopted. Species/OTUs were grouped into sessile opportunistic epibionts, predators/mucous feeders and vagile acrophilic species. The occurrence of each epibiont species and the percentage of colony surface covered were calculated (Table 2). Moreover, recent mechanical damages were also recorded as a percentage of the colony's naked skeleton portion. The total damage was estimated as the sum of the percent of naked portion and epibionted ones (Table 2).

3. Results

3.1. Distribution and Occurrence

The population of *Eunicella verrucosa* of Tavolara MPA is composed of light pinkish-colonies settled on granitic outcrops arising from the detritic bottom and surroundings in the centre of the Tavolara Channel (Figure 1). These outcrops (Figure 2A–C) were characterised by a high level of sedimentation, scarce development of the crustose coralline algae and by the widespread presence of the brown algae *Carpomitra costata*, and to a lesser extent, *Ericaria zosteroides*. The animal community was mainly composed of large, erect sponges, particularly *Axinella* spp. and several species of Keratosa (*Dysidea* spp., *Sarcotragus foetidus*, *Spongia lamella* and *S. officinalis*) (Figure 2D–E). Together with *E. verrucosa*, *Paramuricea clavata* was also relatively abundant.

Colonies (Figure 2F–H) were recorded in 34 of the 77 investigated sites, in an area of approximately 30 ha (Table 1, Figure 1). All the colonies (100) were found between a depth of 35 and 59 m; about one-third of the colonies were settled near the areas where the detritic sediment borders the rocks, whereas the remaining specimens were mainly observed on the sloping flanks of the outcrops and less frequently on their top.

Generally, the colonies were isolated or spread, without the formation of a true forest (sensu Chimienti [28]). The highest number of colonies were found at sites N159 (12 colonies) and N27 (14 colonies), exactly in the middle of the channel (Figure 1).

The colony size of the 100 recorded specimens ranged from 10 to 66.3 cm in height, with the size class 30–40 cm being dominant. Size–frequency distribution was simmetric and leptokurtic (Figure 3A). The fan surface measured for 63 colonies ranged from 27 to 2350 cm^2 and was linearly related to height ($n = 63$; r = 0.73; $p < 0.001$) (Figure 3B).

According to the equation proposed by [39], the age estimation ranges were from 3 to 130 years, with a distribution showing a mode in the 30–50-year-old class, showing a highly skewed and leptokurtic distribution, with a long tail toward large age classes (Figure 3C).

3.2. Epibiosis and Damages

A total of 55% of the observed colonies did not show epibionts or direct damages, whereas the remaining was affected at different levels and in different portions of the colony (base, fan surface, apexes) (Table 2, Figure 4A); no relationship between damages and colony height was observed (on average, healthy colonies were 33.99 ± 1.74 cm height, whereas damaged ones measured 37.22 ± 1.93 cm). Damaged specimens were randomly located across the investigated sites.

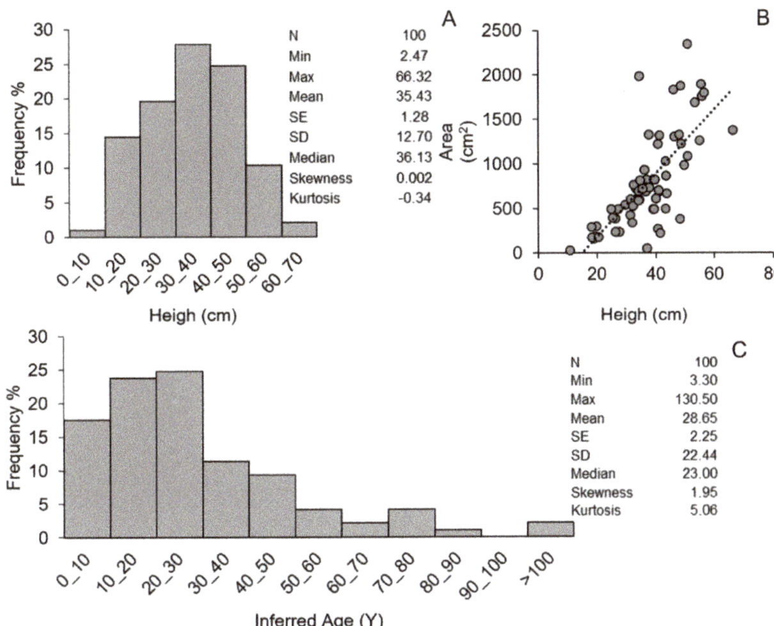

Figure 3. Morphometric description of *Eunicella verrucosa* colonies at the Tavolara MPA. (**A**) size–frequency distribution of the colony heights; (**B**) correlation between height and fan area; (**C**) age–frequency distribution, inferred from height of colonies according to [39].

Only one colony was observed as dead, at site N25 (Figure 5A). Seven colonies showed parts of branches deprived of coenenchyme with an exposed naked skeleton without epibiosis (Figure 5B). In total, 43 colonies hosted epibionts; 29% of these were covered for less than 25% of the total surface, 8% were affected between 25–50%, 4% showed 50–75% of the surface covered and 3% for more than 75% (Figure 4A). Two colonies were recorded entangled by an abandoned nylon line (N151-2 and N159-7) (Figure 5C) and one colony was enveloped by plastic debris (N27-12). This colony was also found spawning on 10 November 2019 (Figure 5D).

In total, 18 species/OTUs were found associated with the colonies (Figure 4B). The most common taxon was the parasitic octocoral *Alcyonium coralloides*, recorded on 39.5% of the damaged colonies (Figure 5E). Overall, branched bryozoans were settled on 67% of the suffering colonies: the most common one was *Turbicellepora avicularis*, present on 37.2% of the colonies (Figure 5F), followed by *Adeonella calveti* and *Pentapora fascialis* (32.6 and 14%, respectively) (Figure 5G). Sponges, particularly *Crella elegans* (25.6%) (Figure 5G), were responsible for the epibiosis on 30% of the colonies. The serpulids of the *Salmacina/Filograna* complex (Figure 5H) and the bivalve *Pteria hirundo* (Figure 5I) were found on 11.6% and 5% of the colonies, respectively.

Three predators, the nudibranch *Duvaucelia odhneri*, the ovulid *Simnia spelta* and the decapod *Balssia gasti* were observed (Figure 6A–C). The most represented was *D. odhneri*, recorded on eleven colonies, in four cases together with their eggs (Figure 6A,A'). Three specimens of *S. spelta* were recorded on two colonies (Figure 6B). One colony (N182-1) hosted ten specimens of *B. gasti* (Figure 6C,C'). Moreover, two specimens of the decapod *Periclimenes scriptus* were recorded on two colonies at site N27 (Figure 6D). Five colonies hosted the large acrophilic ophiuroid, *Astrospartus mediterraneus* (Figure 6E).

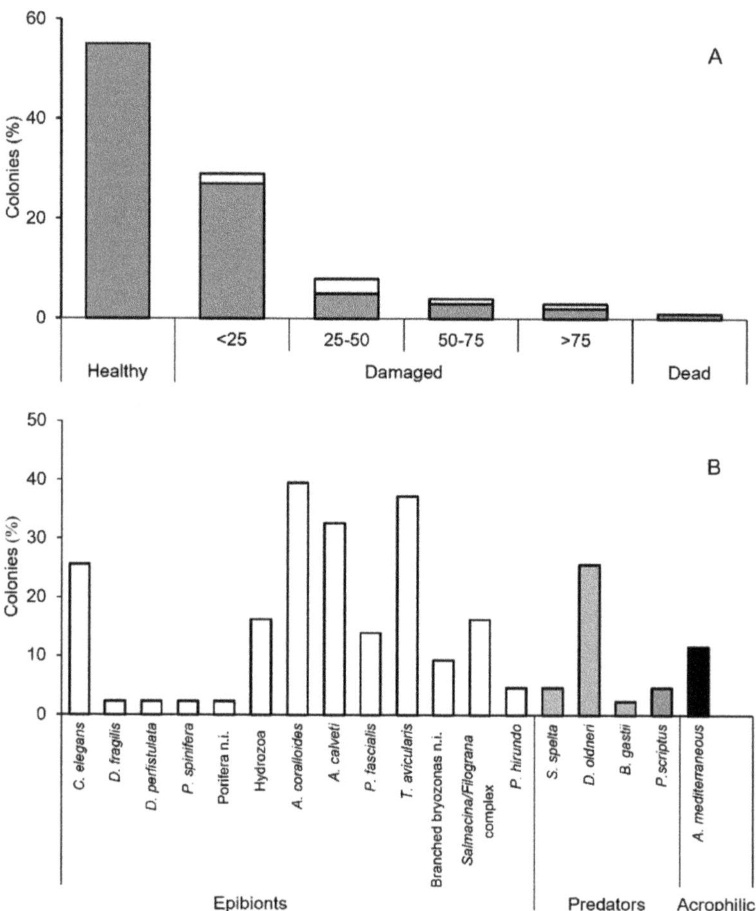

Figure 4. Health state of the studied *Eunicella verrucosa* population. (**A**) percentage of healthy and damaged colonies, according to the percentage of affected surface. White bars, percentage of colonies characterized by naked skeleton; grey bars, percentage of surface covered by epibionts. (**B**) percentage of colonies hosting associated species/OTUs.

Figure 5. Examples of mechanical damages and epibiosis affecting *E. verrucosa*. (**A**) dead colony (white arrow); (**B**) a colony with a huge portion deprived by coenenchyme; (**C**) colonies entangled with an abandoned line and plastic debris (**D**) Red circle refers to spawning polyps; (**E–H**) main epibionts affecting *E. verrucosa*: the parasitic octocoral *Alcyonium coralloides* (**E**) the bryozoans *Turbicellepora avicualaris*, (**F**) *Adeonella calveti*, with the sponge *Crella elegans*, (**G**) the *Salmacina/Filograna* complex (**H**) and (**I**) the bivalve *Pteria hirundo*.

Figure 6. Associated vagile fauna with *Eunicella verrucosa*. The three predators, the nudibranch *Duvaucelia odhneri* (red circles) (**A**) together with its eggs (white circles) and damaged zones (blue arrows), (**A'**) the ovulid *Simnia spelta* (red arrow), (**B**) and the decapod *Balssia gasti*, (**C,C'**) the decapod *Periclimenes scriptus*, (**D**) the large acrophilic ophiuroid and (**E**) *Astrospartus mediterraneous*.

4. Discussion

4.1. Distribution of Eunicella verrucosa and Population Singularity

Although recently the Sardinian coasts have been widely explored through a series of ROV surveys [40–44], no colonies of *Eunicella verrucosa* have been recorded. This evidence agrees with the recent map of the species distribution at the Mediterranean basin scale published by Chimienti [28], which includes data from original investigations, scientific literature and citizen observations validated by photographs.

The record of a persistent population settled in the Tavolara area for at least several decades, is, in this light, a peculiar feature of this zone. Its occurrence in the TPCCMPA was already observed in some previous investigations [29–33].

Here, the *E. verrucosa* colonies settle within sponge-dominated assemblages present on granitic outcrops under a high sedimentation rate in the Tavolara Channel, where the development of crustose coralline algae is limited. A similar assemblage including *E. verrucosa* together with *Axinella polypoides* and massive sponges was described on silted rocks of various lithology at 40–70 m depth in many sites along all Ligurian Sea [8,28].

Our observations indicate that the *E. verrucosa* population is composed of scattered colonies without formation of true forests; in fact, the recorded 100 colonies were grouped on 34 rocky outcrops reaching a total area of about 2 ha. Nevertheless, the size structure of the population seems equilibrated, with a symmetric size–frequency distribution and the modal class of the distribution, 30–40 cm, completely overlapped with that recorded by Chimienti [28] for the denser forests of Sanremo (Ligurian Sea). The estimated age structure reflects previous data obtained in the Marseille region. The data recorded for Sanremo population was younger (11–15 years), with a modal class in the range of 26–30 cm and a tail of old colonies reaching a maximal estimated age of about 71–75 years [28]. It is probable that sexual reproduction is only possible in a cluster of outcrops very close to the central of the channel. The maximal settling distance for this center was about 1.2 km for colonies recorded during this study and about 2 km for the colonies settled on the Klearchos wreck [30]. These data agree with the behavior of the lecithotrophic larvae of *E. verrucosa* showing a dispersion ability around the parent colonies <1 km [5,19,21,45].

The occurrence of several, well-developed colonies on the hull of the Klearchos wreck, sunk on 20 July 1979 [30], is a useful opportunity to validate the age estimation of the population. Although no reference scale is present in the available images, recorded in September 2011, the age of the wreck is in accordance with the modal class of the age distribution recorded during our survey.

The overall rarity of this species in Sardinia is difficult to argue, in the light of the wide occurrence of all the other shallow-water and mesophotic species of alcyonaceans [40–44].

The map of distribution proposed by Chimienti [28] suggests that the predominant Mediterranean water circulation explains the gradual colonization of the western Mediterranean Sea by *E. verrucosa* from the Atlantic Ocean. The presence of this coral in both the Balearic Sea and the Strait of Sicily may be explained by some common environmental and oceanographic features of these two areas. Both are characterized by an intense geostrophic circulation of water masses and a complex seafloor topography, that, due to the presence of islands and seamounts, generates mesoscale eddies and convergent fronts [46–51]. In the Balearic Sea, the colonisation is driven by the ascending Atlantic Water (AW, surface water of Atlantic origin), which, bordering the western coast of Corse, where populations were recorded, see [28], enters the Ligurian Sea, reaching the Tuscany Archipelago. In this area, coral larvae can be spread by the Lyon Gyre [52].

On the other hand, from the Sicily Channel, the species has colonised the Tyrrhenian coast without going beyond the Gulf of Naples. Therefore, the species appears absent in the central Tyrrhenian Sea and the population of Tavolara MPA could be the unique description for the entire sub-basin. In this situation, it is plausible that the occurrence of this species at Tavolara could result from a stochastic event of settling of larvae presumably coming from the Tuscany Archipelago. Genetic studies on this isolated population might help to clarify its origin and connectivity with other coastal forests.

4.2. Predators and Acrophilic Epibionts

The study of the associated community provides some data about the specialized predators of *E. verrucosa*. The tritoniid nudibranch *Duvaucelia odhneri* lives its entire life cycle on the same host colony, exploiting seven different gorgonian species [53], including

E. verrucosa, as also confirmed in this investigation with the discovery of individuals and eggs on the same colony.

The ovulid *Simnia spelta* shows a similar life strategy, being associated with at least four gorgonian species, *Eunicella cavolini*, *E. singularis*, *L. sarmentosa*, and, in our case, *E. verrucosa*. *S. spelta* feeds on the coenenchyme and polyps of the host and also lays ovarian capsules on the branches, causing necrosis of the underlying tissue [54].

During this study, we observed one colony of *E. verrucosa* hosting a group of the palaemonid shrimp *Balssia gasti*. The species has always been observed associated with octocorals, although the nature of the association is still to be elucidated. However, a predatory strategy was hypothesized due to the homocromic camouflage of this species in agreement with the color of the coenenchyme of the hosts [55–58]. In the Tavolara area, *B. gasti* was already observed on *Paramuricea clavata* and *E. cavolini* [59]. Finally, two specimens of the decapod *Periclimenes scriptus*, known as mucus-feeder of octocorals [56], were recorded. In the Tavolara area, this species was observed mainly associated with *P. clavata*, living on the granitic outcrops of the Tavolara Channel [60].

Regarding acrophilic species, the frequent occurrence of the basket star *Astrospartus mediterraneus* is remarkable. This species, generally recorded as colonizing deep habitats, [61] is becoming more and more abundant in relatively shallow waters in recent years [62]. *A. mediterraneus* is one of the few species that have changed its bathymetric distribution moving towards the surface. This is unusual; in fact, in relation to water temperature increasing, numerous shallow-water species changed their bathymetric distribution, reaching deeper levels [63].

4.3. Epibiosis and Health Status of the Population

Out of 100 colonies, 45 showed damages of various entities, from small portions of naked skeleton to completely dead colonies (Figures 5 and 6). The main stressors able to influence the coenenchyme integrity of structuring anthozoans are thermal stress and mechanical injuries, mainly due to fishing activity [34,35,64–66].

Diseases due to heating phenomena with consequent necrosis of coenenchyme are widely documented in the TPCCAMPA for *Paramuricea clavata* and *Eunicella cavolini*, but not at depths exceeding 35–40 m, where the process of necrosis starts from the apical portions of the colonies [67]. Based on this evidence, thermal stress has a negligible influence on the damages described in this study. In fact, the population of *E. verrucosa* mainly lives below the depth where these phenomena are usually documented. Moreover, the degenerative processes due to thermal anomalies are evident as a naked portion of the apical branches. This kind of damage was only sporadically recorded in our study. The direct observation of lost lines entangled in the colonies suggests, at least partially, an anthropic involvement. Physical contact with fishing gear scrapes the gorgonian coenenchyme, favoring the development of epibionts. Epibionts substantially modify the host–environment interactions (e.g., transference of energy or matter), eventually reducing their fitness [68]. Large masses of epibionts lead to a burdening of the colonies and greater mechanical stress, increasing their resistance to water movement [26,34,35,69–71].

The most common epibionts observed in this study were *Alcyonium coralloides*, bryozoans (*Turbicellepora avicularis*, *Adeonella calveti* and *Pentapora fascialis*) and the demosponge *Crella elegans* (Figures 4B and 5).

A. coralloides is one of the first colonizers of the skeletal portions deprived of coenenchyme and can subsequently continue its expansion to the detriment of the coral to fully occupy its skeleton. On the Tavolara specimens, *A. coralloides* always settled in the basal or central portions of the fan and never on the apical parts. It is generally accepted that the high occurrence of this epibiont may be correlated, in frequented sites, to anthropogenic damages [72], supporting its use as a bioindicator of stress in coralligenous assemblages [26,34,73,74]. Similar considerations are also true for erect bryozoans, such as *T. avicularis* [75].

Many Mediterranean localities endure impacts by anthropogenic pressure due to demersal fishing activities that pauperize three-dimensional benthic ecosystems, such as coral forests [10,26,35,40,69,76–81]. In the Medes Islands (Catalan Sea), between 10% and 33% of the colonies in unprotected populations were partially colonized by epibionts, most likely following tissue injury, whereas only from 4% to 10% of the populations in a marine protected area was affected [76], suggesting that fishing activities directly cause severe damage expressed as epibiosis coverage. Our data indicate an epibiosis at least four times higher for the Tavolara's *E. verrucosa* population.

The communities occurring on the granite outcrops have traditionally been considered of low quality due to the absence of the typical coralligenous features such as the thick coralline algal concretion [82]. This underestimation serves to classify the Tavolara Channel in the C-zone of MPA (Partial Reserve), allowing artisanal and recreational fishing activities, which, in turn, probably increased the pressure on the benthic communities. Recently, however, it was stated that the communities settled in this particular habitat are not impoverished facies of the coralligenous assemblage, but a peculiar community composed of erect sponge and habitat-forming anthozoans [33]. An adjustment of the management guidelines of the MPA is required in light of the re-evaluation of this habitat.

Author Contributions: Conceptualization, G.B., M.C. and E.T.; methodology, M.C. and E.T.; validation, M.C., M.B. and F.E.; formal analysis, M.C.; investigation, M.C. and E.T.; resources, E.T.; data curation, E.T.; writing—original draft preparation, M.C., E.T., M.B., F.E. and G.B.; writing—review and editing, M.C., E.T., M.B., F.E. and G.B.; visualization, M.C. and E.T.; supervision, M.B. and G.B.; project administration, G.B. All authors have read and agreed to the published version of the manuscript.

Funding: This research received no external funding.

Institutional Review Board Statement: Not applicable.

Informed Consent Statement: Not applicable.

Data Availability Statement: The data is available upon request form a corresponding auhor.

Acknowledgments: The authors would like to thank the Tavolara MPA management for the permission to use the GIS environment and for the possibility to independently develop the underwater surveys. The authors would like to thank the "Slow dive" team for its support during the diving activities. Special thanks go to Florian Holon and Laurent Ballesta of ANDROMEDE Oceanologie, for providing the photographic material on the Klearchos wreck.

Conflicts of Interest: The authors declare no conflict of interest.

References

1. Carpine, C. Les Gorgonaries de la Mediterranée. *Bull. Musee Oceanogr. Monaco* **1975**, *71*, 1–140.
2. Manuel, R.L. *British Anthozoa (Coelenterata: Octocorallia and Hexacorallia)*; Estuarine and Brackish-Water Sciences Association: Leiden, The Netherlands, 1988.
3. Readman, J.A.J.; Hiscock, K. Eunicella verrucosa Pink sea fan. In *Marine Life Information Network: Biology and Sensitivity Key Information Reviews*; Tyler-Walters, H., Hiscock, K., Eds.; Marine Biological Association of the United Kingdom: Plymouth, UK, 2017; Available online: https://www.marlin.ac.uk/species/detail/1121 (accessed on 9 February 2022).
4. Lafargue, F. Peuplements sessiles de l'Archipel de Glénan L'Inventaire: Anthozoaires. *Vie Milieu* **1969**, *20*, 415–436.
5. Hiscock, K. Changes in the marine life of Lundy. *Rep. Lundy Field Soc.* **2003**, *53*, 86–95.
6. Sartoretto, S.; Francour, P. Bathymetric distribution and growth rates of *Eunicella verrucosa* (Cnidaria: Gorgoniidae) populationsalong the Marseilles coast (France). *Sci. Mar.* **2012**, *76*, 349–355. [CrossRef]
7. Di Camillo, C.G.; Ponti, M.; Bavestrello, G.; Krzelj, M.; Cerrano, C. Building a baseline for habitat-forming corals by a multi-source approach, including Web Ecological Knowledge. *Biodivers. Conserv.* **2018**, *27*, 1257–1276. [CrossRef]
8. Enrichetti, F.; Dominguez-Carrió, C.; Toma, M.; Bavestrello, G.; Betti, F.; Canese, S.; Bo, M. Megabenthic communities of the Ligurian deep continental shelf and shelf break (NW Mediterranean Sea). *PLoS ONE* **2019**, *14*, e0223949. [CrossRef]
9. IUCN (International Union for Conservation of Nature and Natural Resources). IUCN Red List of Threatened Species. 2022. Available online: www.iucnredlist (accessed on 9 February 2022).

10. Grinyó, J.; Gori, A.; Ambroso, S.; Purroy, A.; Calatayud, C.; Dominguez-Carrió, C.; Coppari, M.; Lo Iacono, C.; López-González, P.J.; Gili, J.M. Diversity, distribution and population size structure of deep Mediterranean gorgonian assemblages (Menorca Channel, Western Mediterranean Sea). *Progr. Oceanogr.* **2016**, *145*, 42–56. [CrossRef]

11. Chimienti, G.; Stithou, M.; Dalle Mura, I.; Mastrototaro, F.; D'Onghia, G.; Tursi, A.; Izzi, C.; Fraschetti, S. An explorative assessment of the importance of Mediterranean Coralligenous habitat to local economy: The case of recreational diving. *J. Environ. Account. Manag.* **2017**, *5*, 315–325. [CrossRef]

12. Velimirov, B. Orientation in the sea fan Eunicella cavolinii related to water movement. *Helgol. Wiss. Meeresunters.* **1973**, *24*, 163–173. [CrossRef]

13. Velimirov, B. Variations in growth forms of Eunicella cavolini Koch (Octocorallia) related to intensity of water movement. *J. Exp. Mar. Biol. Ecol.* **1976**, *21*, 109–117. [CrossRef]

14. West, J.M. Plasticity in the sclerites of a gorgonian coral: Tests of water motion, light level, and damage cues. *Biol. Bull.* **1997**, *192*, 279–289. [CrossRef] [PubMed]

15. Weinbauer, M.G.; Velimirov, B. Morphological variations in the Mediterranean sea fan *Eunicella cavolini* (Coelenterata: Gorgonacea) in relation to exposure, colony size and colony region. *Bull. Mar. Sci.* **1995**, *56*, 283–295.

16. Kim, K.; Lasker, H.R. Allometry of resource capture in colonial cnidarians and constraints on modular growth. *Funct. Ecol.* **1998**, *12*, 646–654. [CrossRef]

17. Skoufas, G. Comparative biometry of Eunicella singularis (Gorgonian) sclerites at East Mediterranean Sea (North Aegean Sea, Greece). *Mar. Biol.* **2006**, *149*, 1365–1370. [CrossRef]

18. Chang, W.L.; Chi, K.J.; Fan, T.Y.; Dai, C.F. Skeletal modification in response to flow during growth in colonies of the sea whip, Junceella fragilis. *J. Exp. Mar. Biol. Ecol.* **2007**, *347*, 97–108. [CrossRef]

19. Munro, L.; Munro, C. Reef Research. *Determining the Reproductive Cycle of Eunicella verrucosa*; Interim report March 2003. RR Report 10 Nov. 2003. Available online: http://www.marine-bio-images.com/RR_Eunicella_PDFS/Report_RR12Jul2004reproductive%20cycle%20pdf.pdf (accessed on 9 February 2022).

20. Goffredo, S.; Lasker, H.R. Modular growth of a gorgonian coral can generate predictable patterns of colony growth. *J. Exp. Mar. Biol. Ecol.* **2006**, *336*, 221–229. [CrossRef]

21. Hiscock, K. *Eunicella verrucosa. Pink Sea Fan. Marine Life Information Network: Biology and Sensitivity Key Information Sub-programme.* 2007. Available online: https://www.marlin.ac.uk/assets/pdf/species/marlin_species_1121_2019-03-21.pdf (accessed on 9 February 2022).

22. Coz, R.; Ouisse, V.; Artero, C.; Carpentier, A.; Crave, A.; Feunteun, E.; Olivier, J.M.; Perrin, B.; Ysnel, F. Development of a new standardised method for sustainable monitoring of the vulnerable pink sea fan *Eunicella verrucosa*. *Mar. Biol.* **2012**, *159*, 1375–1388. [CrossRef]

23. Coma, R.; Pola, E.; Ribes, M.; Zabala, M. Long-term assessment of temperate octocoral mortality patterns, protected vs. unprotected areas. *Ecol. Appl.* **2004**, *14*, 1466–1478. [CrossRef]

24. Linares, C.; Coma, R.; Zabala, M. Restoration of threatened red gorgonian populations: An experimental and modelling approach. *Biol. Conserv.* **2008**, *141*, 427–437. [CrossRef]

25. Lloret, J.; Zaragoza, N.; Caballero, D.; Riera, V. Impacts of recreational boating on the marine environment of Cap de Creus (Mediterranean Sea). *Ocean Coast. Manag.* **2008**, *51*, 749–754. [CrossRef]

26. Enrichetti, F.; Bava, S.; Bavestrello, G.; Betti, F.; Lanteri, L.; Bo, M. Artisanal fishing impact on deep coralligenous animal forests: A Mediterranean case study of marine vulnerability. *Ocean Coast. Manag.* **2019**, *177*, 112–126. [CrossRef]

27. Otero, M.M.; Numa, C.; Bo, M.; Orejas, C.; Garrabou, J.; Cerrano, C.; Kružić, P.; Antoniadou, C.; Aguilar, R.; Kipson, S.; et al. *Overview of the Conservation Status of Mediterranean Anthozoans*; IUCN: Málaga, Spain, 2017; pp. 1–73.

28. Chimienti, G. Vulnerable Forests of the Pink Sea Fan Eunicella verrucosa in the Mediterranean Sea. *Diversity* **2020**, *12*, 176. [CrossRef]

29. Bianchi, C.N.; Morri, C.; Navone, A. I popolamenti delle scogliere rocciose sommerse dell'Area Marina Protetta di Tavolara Punta Coda Cavallo (Sardegna nord-orientale). *Sci. Rep. Port Cros Natl. Park* **2014**, *24*, 39–85.

30. ANDROMEDE. *Etude et Cartographie des Biocénoses Marines de l'Aire Marine Protégée de Tavolara—Punta Coda Cavallo, Sardaigne (Italie)*; Initiative pour les Petites Iles de Méditerranée; Contrat Œil d' Andromède/Agence de L'Eau: Lion, France, 2012; p. 149.

31. ANDROMEDE. *Inventaire et Cartographie des Assemblages Coralligènes de l'Aire Marine Protégée de Tavolara—Punta Coda Cavallo, Sardaigne (Italie)*; Initiative pour les Petites Iles de Méditerranée; Andromède Océanologie/Agence de l'Eau: Lion, France, 2016; p. 100.

32. Canessa, M.; Bavestrello, G.; Bo, M.; Trainito, E.; Panzalis, P.; Navone Caragnano, A.; Betti, F.; Cattaneo-Vietti, R. Coralligenous assemblages differ between limestone and granite: A case study at the Tavolara-Punta Coda Cavallo Marine Protected Area (NE Sardinia, Mediterranean Sea). *Reg. Stud. Mar. Sci.* **2020**, *35*, 101159. [CrossRef]

33. Canessa, M.; Bavestrello, G.; Trainito, E.; Bianchi, C.N.; Morri, C.; Navone, A.; Cattaneo-Vietti, R. A large and erected sponge assemblage on granite outcrops in a Mediterranean Marine Protected Area (NE Sardinia). *Reg. Stud. Mar. Sci.* **2021**, *44*, 101734. [CrossRef]

34. Bavestrello, G.; Cerrano, C.; Zanzi, D.; Cattaneo-Vietti, R. Damage by fishing activities to the Gorgonian coral *Paramuricea clavata* in the Ligurian Sea. *Aquat. Conserv.* **1997**, *7*, 253–262. [CrossRef]

35. Bo, M.; Bava, S.; Canese, S.; Angiolillo, M.; Cattaneo-Vietti, R.; Bavestrello, G. Fishing impact on deep Mediterranean rocky habitats as revealed by ROV investigation. *Biol. Conserv.* **2014**, *171*, 167–176. [CrossRef]

36. Pititto, F.; Trainito, E.; Macic, V.; Rais, C.; Torchia, G. The resolution in benthic carthography: A detailed mapping technique and a multiscale GIS approach with applications to coralligenous assemblages. In Proceedings of the Conference: 2° Mediterranean Symposium on the Conservation of Coralligenous & Other Calcareous Bio-Concretions, Portorož, Slovenia, 29–30 October 2014. [CrossRef]

37. Rasband, W.S. *ImageJ: Image Processing and Analysis in Java*; ascl-1206; National Institutes of Health: Bethesda, MD, USA, 2012.

38. Coma, R.; Ribes, M.; Zabala, M.; Gili, J.M. Growth in a modular colonial marine invertebrate. *Estuar. Coast. Shelf Sci.* **1998**, *47*, 459–470. [CrossRef]

39. Chimienti, G.; Di Nisio, A.; Lanzolla, A.M.L. Size/Age Models for Monitoring of the Pink Sea Fan *Eunicella verrucosa* (Cnidaria: Alcyonacea) and a Case Study Application. *J. Mar. Sci. Eng.* **2020**, *8*, 951. [CrossRef]

40. Bo, M.; Bavestrello, G.; Angiolillo, M.; Calcagnile, L.; Canese, S.; Cannas, R.; Cau, A.; D'Elia, M.; D'Oriano, F.; Follesa, M.C.; et al. Persistence of pristine deep-sea coral gardens in the Mediterranean Sea (SW Sardinia). *PLoS ONE* **2015**, *10*, e0119393. [CrossRef]

41. Cau, A.; Follesa, M.C.; Moccia, D.; Alvito, A.; Bo, M.; Angiolillo, M.; Canese, S.; Paliaga, E.M.; Orrù, P.E.; Sacco, F.; et al. Deepwater corals biodiversity along roche du large ecosystems with different habitat complexity along the south Sardinia continental margin (CW Mediterranean Sea). *Mar. Biol.* **2015**, *162*, 1865–1878. [CrossRef]

42. Gori, A.; Grinyó, J.; Dominguez-Carrió, C.; Ambroso, S.; López-González, P.J.; Gili, J.M.; Bavestrello, G.; Bo, M. 20 Gorgonian and black coral assemblages in deep coastal bottoms and continental shelves of the Mediterranean Sea. In *Mediterranean Cold-Water Corals: Past, Present and Future*; Springer: Cham, Switzerland, 2019; pp. 245–248.

43. Moccia, D.; Alvito, A.; Follesa, M.C. Preliminary ROV surveys data on deep-coral assemblages from South-East Sardinian Sea (central western Mediterranean). In Proceedings of the Marine Imaging Workshop, Southampton, UK, 7–10 April 2014.

44. Moccia, D.; Cau, A.; Bramanti, L.; Carugati, L.; Canese, S.; Follesa, M.C.; Cannas, R. Spatial distribution and habitat characterization of marine animal forest assemblages along nine submarine canyons of Eastern Sardinia (central Mediterranean Sea). *Deep. Sea Res. Part I Oceanogr. Res. Pap.* **2021**, *167*, 103422. [CrossRef]

45. Weinberg, S.; Weinberg, F. The life cycle of a gorgonian: Eunicella singularis (Esper, 1794). *Bijdr. Dierkd.* **1979**, *48*, 127–140. [CrossRef]

46. López Jurado, L.F.; González Barbuzano, J.; Hildebrandt, S. *La foca Monje y las Islas Canarias: Biología, Ecología y Conservación de una Especie*; Consejería de Política Territorial, Gobierno de Canarias: Las Palmas de Gran Canaria, Spain, 1995; ISDN 8460623610.

47. Pinot, J.M.; Tintoré, J.; Gomis, D. Multivariate analysis of the surface circulation in the Balearic Sea. *Progr. Oceanogr.* **1995**, *36*, 343–376. [CrossRef]

48. Vélez-Belchí, P.; Tintoré, J. Vertical velocities at an ocean front. *Sci. Mar.* **2001**, *65* (Suppl. S1), 291–300. [CrossRef]

49. García, A.; Alemany, F.; Vélez-Belchí, P.; López Jurado, J.L.; Cortés, D.; de la Serna, J.M.; González Pola, C.; Rodríguez, J.M.; Jansá, J.; Ramírez, T. Characterization of the bluefin tuna spawning habitat off the Balearic Archipelago in relation to key hydrographic features and associated environmental conditions. *Collect. Vol. Sci. Pap. ICCAT* **2003**, *58*, 535–549.

50. Oray, I.; Karakulak, S.; Garcia, A.; Piccinetti, C.; Rollandi, L.; de la Serna, J.M. Report on the Mediterranean BYP tuna larval meeting. *Coll. Vol. Sci. Pap. ICCAT* **2005**, *58*, 1429–1435.

51. Alemany, F.; Deudero, S.; Morales-Nin, B.; Lopez-Jurado, J.L.; Jansa, J.; Palmer, M.; Palomera, I. Influence of physical environmental factors on the composition and horizontal distribution of summer larval fish assemblages off Mallorca Island (Balearic archipelago, Western Mediterranean). *J. Plankton Res.* **2006**, *28*, 473–487. [CrossRef]

52. Vergnaud-Grazzini, C.; Borsetti, A.M.; Cati, F.; Colantoni, P.; D'Onofrio, S.; Saliege, J.F.; Sartori, R.; Tampieri, R. Palaeoceanographic record of the last deglaciation in the Strait of Sicily. *Mar. Micropaleontol.* **1988**, *13*, 1–21. [CrossRef]

53. Furfaro, G.; Trainito, E.; De Lorenzi, F.; Fantin, M.; Doneddu, M. *Tritonia nilsodhneri* Marcus Ev., 1983 (Gastropoda, Heterobranchia, Tritoniidae): First records for the Adriatic Sea and new data on ecology and distribution of Mediterranean populations. *Acta Adriat.* **2017**, *58*, 261–270. [CrossRef]

54. Theodor, J. Contribution a l'étude des gorgones (VI): La dénudation des branches de gorgones par des mollusques prédateurs. *Vie Milieu* **1967**, *18*, 73–78.

55. Manconi, R.; Mori, M. New records of *Balssia gasti* (BALSS, 1921) (Decapoda, Palaemonidae) in the western Mediterranean Sea. *Crustaceana* **1990**, *59*, 96–100. [CrossRef]

56. Manconi, R.; Mori, M. Caridean shrimps (Decapoda) found among *Corallium rubrum* (L., 1758). *Crustaceana* **1992**, *62*, 105–110. [CrossRef]

57. Mori, M.; Morri, C.; Bianchi, C.N. Notes on the life history of the pontonine shrimp *Balssia gasti* (Balss, 1921). *Oebalia* **1994**, *20*, 129–137.

58. Trainito, E.; Baldacconi, R. *Atlante di Flora e Fauna del Mediterraneo*; Il Castello Editore: Cornaredo, Italy, 2021; pp. 262–448.

59. Ponti, M.; Grech, D.; Mori, M.; Perlini, R.A.; Ventra, V.; Panzalis, P.A.; Cerrano, C. The role of gorgonians on the diversity of vagile benthic fauna in Mediterranean rocky habitats. *Mar. Biol.* **2016**, *163*, 120. [CrossRef]

60. Canessa, M.; Bavestrello, G.; Guidetti, P.; Navone, A.; Trainito, E. Marine rocky reef assemblages and lithological properties of substrates are connected at different ecological levels. *Eur. Zool. J.* **2022**, in press.

61. Zibrowius, H. Observations biologiques au large du Lavandou (côte méditerranéenne de France) à l'aide du sous-marin Griffon de la Marine Nationale. *Trav. Sci. Parc. Natl. Port Cros* **1978**, *4*, 171–176.

62. Montasell Bartrés, M. *Astrospartus mediterraneus* (Echinodermata: Ophiuroidea) in the Cap de Creus Marine Area: Ecological Characterization of an Emblematic Species. Bachelor's Thesis, Universitat Central de Catalunya, Vic, Spain, 2020.
63. Puce, S.; Bavestrello, G.; Di Camillo, C.G.; Boero, F. Long-term changes in hydroid (Cnidaria, Hydrozoa) assemblages: Effect of Mediterranean warming? *Mar. Ecol.* **2009**, *30*, 313–326. [CrossRef]
64. Bavestrello, G.; Bertone, S.; Cattaneo-Vietti, R.; Cerrano, C.; Gaino, E.; Zanzi, D. Mass mortality of *Paramuricea clavata* (Anthozoa, Cnidaria) on Portofino Promontory cliffs, Ligurian Sea, Mediterranean Sea. *Mar. Life* **1994**, *4*, 15–19.
65. Cerrano, C.; Bavestrello, G. Medium-term effects of die-off of rocky benthos in the Ligurian Sea. What can we learn from gorgonians? *Chem. Ecol.* **2008**, *24* (Suppl. S1), 73–82. [CrossRef]
66. Torrents, O.; Tambutté, E.; Caminiti, N.; Garrabou, J. Upper thermal thresholds of shallow vs. deep populations of the precious Mediterranean red coral *Corallium rubrum* (L.): Assessing the potential effects of warming in the NW Mediterranean. *J. Exp. Mar. Biol. Ecol.* **2008**, *357*, 7–19. [CrossRef]
67. Huete-Stauffer, C.; Vielmini, I.; Palma, M.; Navone, A.; Panzalis, P.; Vezzulli, L.; Misic, C.; Cerrano, C. *Paramuricea clavata* (Anthozoa, Octocorallia) loss in the Marine Protected Area of Tavolara (Sardinia, Italy) due to a mass mortality event. *Mar. Ecol.* **2011**, *32*, 107–116. [CrossRef]
68. Wahl, M. Ecological lever and interface ecology: Epibiosis modulates the interactions between host and environment. *Biofouling* **2008**, *24*, 427–438. [CrossRef] [PubMed]
69. Harmelin, J.G.; Marinopoulos, J. Population structure and partial mortality of the gorgonian *Paramuricea clavata* (Risso) in the north-western Mediterranean (France, Port-Cros Island). *Mar. Life* **1994**, *4*, 5–13.
70. Riegl, B.; Riegl, A. Studies on coral community structure and damage as a basis for zoning marine reserves. *Biol. Conserv.* **1996**, *77*, 269–277. [CrossRef]
71. Cerrano, C.; Arillo, A.; Azzini, F.; Calcinai, B.; Castellano, L.; Muti, C.; Valisano, L.; Zega, G.; Bavestrello, G. Gorgonian population recovery after a mass mortality event. *Aquat. Conserv.* **2005**, *15*, 147–157. [CrossRef]
72. Groot, S.; Weinberg, S. Biogeography, taxonomical status and ecology of Alcyonium (Parerythropodium) coralloides (Pallas, 1766). *Mar. Ecol.* **1982**, *3*, 293–312. [CrossRef]
73. Cerrano, C.; Bavestrello, G.; Bianchi, C.N.; Cattaneo-Vietti, R.; Bava, S.; Morganti, C.; Morri, C.; Picco, P.; Sara, G.; Schiaparelli, S.; et al. A catastrophic mass- mortality episode of gorgonians and other organisms in the Ligurian Sea (northwestern Mediterranean), summer 1999. *Ecol. Lett.* **2000**, *3*, 284–293. [CrossRef]
74. Quintanilla, E.; Gili, J.-M.; Lòpez-Gonzàlez, P.J.; Tsounis, G.; Madurell, T.; Fiorillo, I.; Rossi, S. Sexual reproductive cycle of the epibiotic soft coral Alcyonium coralloides (Octocorallia, Alcyonacea). *Aquat. Biol.* **2013**, *18*, 113–124. [CrossRef]
75. Hall-Spencer, J.M.; Pike, J.; Munn, C.B. Diseases affect cold-water corals too: *Eunicella verrucosa* (Cnidaria: Gorgonacea) necrosis in SW England. *Dis. Aquat. Org.* **2007**, *76*, 87–97. [CrossRef]
76. Tsounis, G.; Martinez, L.; Bramanti, L.; Viladrich, N.; Gili, J.M.; Martinez, Á.; Rossi, S. Anthropogenic effects on reproductive effort and allocation of energy reserves in the Mediterranean octocoral *Paramuricea clavata*. *Mar. Ecol. Prog. Ser.* **2012**, *449*, 161–172. [CrossRef]
77. Angiolillo, M.; di Lorenzo, B.; Farcomeni, A.; Bo, M.; Bavestrello, G.; Santangelo, G.; Cau, A.; Mastascusa, V.; Cau, A.; Sacco, F.; et al. Distribution and assessment of marine debris in the deep Tyrrhenian Sea (NW Mediterranean Sea, Italy). *Mar. Pollut. Bull.* **2015**, *92*, 149–159. [CrossRef] [PubMed]
78. Sini, M.; Kipson, S.; Linares, C.; Koutsoubas, D.; Garrabou, J. The yellow gorgonian Eunicella cavolini: Demography and disturbance levels across the Mediterranean Sea. *PLoS ONE* **2015**, *10*, e0126253. [CrossRef] [PubMed]
79. Gori, A.; Bavestrello, G.; Grinyó, J.; Dominguez-Carrió, C.; Ambroso, S.; Bo, M. Animal forests in deep coastal bottoms and continental shelf of the Mediterranean Sea. In *Marine Animal Forests: The Ecology of Benthic Biodiversity Hotspots*; Springer: Berlin/Heidelberg, Germany, 2017; pp. 207–233.
80. D'Onghia, G.; Calculli, C.; Capezzuto, F.; Carlucci, R.; Carluccio, A.; Grehan, A.; Indennidate, P.; Maiorano, F.; Mastrototaro, A.; Pollice, T.; et al. Anthropogenic impact in the Santa Maria di Leuca cold-water coral province (Mediterranean Sea): Observations and conservation straits. *Deep. Sea Res. Part II Top. Stud. Oceanogr.* **2017**, *145*, 87–101. [CrossRef]
81. Enrichetti, F.; Dominguez-Carrió, C.; Toma, M.; Bavestrello, G.; Canese, S.; Bo, M. Assessment and distribution of seafloor litter on the deep Ligurian continental shelf and shelf break (NW Mediterranean Sea). *Mar. Pollut. Bull.* **2020**, *151*, 110872. [CrossRef]
82. Rovere, A.; Ferraris, F.; Parravicini, V.; Navone, A.; Morri, C.; Bianchi, C.N. Characterization and evaluation of a marine protected area: 'Tavolara–Punta Coda Cavallo' (Sardinia, NW Mediterranean). *J. Maps* **2013**, *9*, 279–288. [CrossRef]

diversity

MDPI

Article

Morphological Modifications and Injuries of Corals Caused by Symbiotic Feather Duster Worms (Sabellidae) in the Caribbean

Bert W. Hoeksema [1,2,3,*], Rosalie F. Timmerman [1,2], Roselle Spaargaren [1,2], Annabel Smith-Moorhouse [1,2], Roel J. van der Schoot [1,2], Sean J. Langdon-Down [1,2] and Charlotte E. Harper [1,2]

[1] Taxonomy, Systematics and Geodiversity Group, Naturalis Biodiversity Center, P.O. Box 9517, 2300 RA Leiden, The Netherlands; r.f.timmerman@student.rug.nl (R.F.T.); r.spaargaren@student.rug.nl (R.S.); a.smith-moorhouse@student.rug.nl (A.S.-M.); roel.vanderschoot@naturalis.nl (R.J.v.d.S.); s.langdon-down@student.rug.nl (S.J.L.-D.); c.harper@student.rug.nl (C.E.H.)
[2] Groningen Institute for Evolutionary Life Sciences, University of Groningen, P.O. Box 11103, 9700 CC Groningen, The Netherlands
[3] Institute of Biology Leiden, Leiden University, P.O. Box 9505, 2300 RA Leiden, The Netherlands
* Correspondence: bert.hoeksema@naturalis.nl

Abstract: Some coral-associated invertebrates are known for the negative impact they have on the health of their hosts. During biodiversity surveys on the coral reefs of Curaçao and a study of photo archives of Curaçao, Bonaire, and St. Eustatius, the Caribbean split-crown feather duster worm *Anamobaea* sp. (Sabellidae) was discovered as an associate of 27 stony coral species (Scleractinia spp. and *Millepora* spp.). The worm was also found in association with an encrusting octocoral (*Erythropodium caribaeorum*), a colonial tunicate (*Trididemnum solidum*), various sponge species, and thallose algae (mainly *Lobophora* sp.), each hypothesized to be secondary hosts. The worms were also common on dead coral. Sabellids of the genera *Bispira* and *Sabellastarte* were all found on dead coral. Some of them appeared to have settled next to live corals or on patches of dead coral skeleton surrounded by living coral tissue, forming pseudo-associations. Associated *Anamobaea* worms can cause distinct injuries in most host coral species and morphological deformities in a few of them. Since *Anamobaea* worms can form high densities, they have the potential to become a pest species on Caribbean coral reefs when environmental conditions become more favorable for them.

Keywords: *Anamobaea*; *Bispira*; coral damage; host generalist; Polychaeta; pseudo-association; *Sabellastarte*

Citation: Hoeksema, B.W.; Timmerman, R.F.; Spaargaren, R.; Smith-Moorhouse, A.; van der Schoot, R.J.; Langdon-Down, S.J.; Harper, C.E. Morphological Modifications and Injuries of Corals Caused by Symbiotic Feather Duster Worms (Sabellidae) in the Caribbean. *Diversity* **2022**, *14*, 332. https://doi.org/10.3390/d14050332

Academic Editor: Cinzia Corinaldesi

Received: 27 March 2022
Accepted: 21 April 2022
Published: 25 April 2022

Publisher's Note: MDPI stays neutral with regard to jurisdictional claims in published maps and institutional affiliations.

1. Introduction

As foundation species, reef corals provide a habitat to a large diversity of marine invertebrates, which represent a variety of phyla [1–4]. A large proportion of these invertebrate taxa use these corals as living hosts, whereas others only need dead coral as a rocky substrate for settlement and growth. The first category mostly contains species that live in strict symbiotic relations with their host corals (be it commensalistic, mutualistic, or parasitic) and are generally known as coral-associated fauna [5,6]. Due to their vulnerability to disturbance, their presence is supposed to be indicative of reef health [7,8]. These relations may vary because, in some studies, coral-associated species are reported as beneficial to their host by offering protection against predators and diseases [9–13] or cleaning services [14]. In other hosts, associated species are shown to be harmful by causing coral injuries or by obstructing the host's growth [15–20].

It is not precisely known if some reef-dwelling invertebrates, such as feather duster worms (fan worms) of the family Sabellidae, live in symbiosis with corals. Sabellids are tube-forming, solitary, or colonial sedentary polychaetes occurring in benthic environments. The protective tube is usually flexible and predominantly buried in sediment or attached to a hard substrate [21]. The animals have two sets of colorful radiolar tentacles (radioles), which normally extend from their tube and are used for feeding and respiration [22–24].

Although various sabellid species have been reported to live in coral reefs or more specifically on dead coral [25–32], they have received little or no attention in the literature about coral-associated fauna [8,33–45] and symbiotic polychaetes [46–48], in contrast with serpulid worms. Only a few publications mention the identity of sabellid worms and their host coral species, such as the sabellids *Amphicorina schlenzae* Nogueira & Amaral, 2000 and *Pseudobranchiomma minima* Nogueira & Knight-Jones, 2002 in living colonies of the Brazilian endemic scleractinian *Mussimilia hispida* (Verrill, 1901) [49,50]. Furthermore, there are records from Indonesia of *Perkinsiana anodina* Capa, 2007 in an encrusting mushroom coral *Cycloseris explanulata* (van der Horst, 1922), misidentified as *C. wellsi* (Veron & Pichon, 1980) [51,52], and *Notaulax montiporicola* Tovar-Hernández & ten Hove, 2020, associated with the foliaceous coral *Montipora nodosa* (Dana, 1846) [24,32,51]. Finally, the fan worm *Notaulax yamasui* Nishi et al., 2017 was recorded from dead and living *Porites* sp. in Okinawa Island, southern Japan [53]. None of these association records are from the Caribbean. However, there is a published photograph of a colonial feather duster worm *Bispira brunnea* (Treadwell, 1917) on top of a coral wound of an unidentified scleractinian in the Mexican Caribbean [54].

During a recent biodiversity survey of coral reefs of Curaçao (southern Caribbean), associations of split-crown feather duster worms (*Anamobaea* sp.) [22,27,29] with corals were observed to be abundant. Because these associations were not reported before and the presence of these worms appeared to cause aberrant growth forms and injuries in host corals, we investigated which host coral species were affected. The present report serves to create awareness of these associations and of the potential damage the worms may cause to Caribbean coral reefs. Several sabellids of the genera *Bispira* Krøyer, 1856 and *Sabellastarte* Krøyer, 1856 [22,27,29] were found in close proximity to corals, but appeared to have settled next to their hosts or on patches of dead coral skeleton surrounded by living coral tissue.

2. Materials and Methods

The surveys took place during October–December 2021 and April 2022 along the leeward side of the island of Curaçao at depths down to 20 m. To investigate the preferred habitats of symbiotic feather duster worms, all observed host coral species were recorded and photographed, as well as other host species that were encountered. Because coral-dwelling feather duster worms were not recorded before in the Caribbean, coral photographs taken by the first author during earlier surveys were also checked for the presence of symbiotic feather duster worms: Curaçao (in 2017, 2015, and 2014), Bonaire (in 2019), and St. Eustatius (in 2015). Curaçao and Bonaire are located in the Southern Caribbean, and St. Eustatius is in the Eastern Caribbean (Figure 1). All association records were listed per island and year (Table 1).

3. Results

Twenty-seven host-coral species, consisting of 25 scleractinians (Anthozoa) and two milleporids (Hydrozoa), were recorded for the coral-associated feather duster worm, divided over 10 families and 16 genera (Table 1; Figures 2–7). In addition, the species was found in association with the encrusting octocoral *Erythropodium caribaeorum* (Figure 8A,B), the colonial tunicate *Trididemnum solidum* (Figure 8C,D), phaeophyceaen algae, in particular *Lobophora* sp. (Figure 8E,F), and various sponge species (Figure 9). The records were from the southern Caribbean islands of Bonaire and Curaçao and the Eastern Caribbean island of St. Eustatius (Table 1).

Figure 1. Map of the eastern part of the Caribbean showing the position of Curaçao, Bonaire, and St. Eustatius, where the presence of coral-associated feather duster worms was investigated.

The symbiotic worms, identified as split-crown feather duster worms of the genus *Anamobaea* Krøyer, 1856 [22,27,29], showed some variation in coloration, ranging from white to dark red and various combination patterns of these colors (Figures 2, 3 and 9A). Two species from the Caribbean have been described, which can be distinguished by two morphological characters [27,29,32] that are not clearly visible in the photographs: *Anamobaea phyllisae* Tovar-Hernández & Salazar-Vallejo, 2006 has two dorsal kidney-shaped shields over the anterior margin of the base of its crown and smooth flanges (without papillae) and *Anamobaea orstedi* Krøyer, 1856 does not have such shields, and its flanges are wrinkled (with papillae). The former species has so far only been reported from the type locality in the British Virgin Islands, whereas the latter has a wider geographic range [29,32]. Because we are not sure about the identity of the associated worms, we refer to them as *Anamobaea* sp.

Most observed worms were withdrawn in their tubes; only a few of them were observed with extended radioles protruding from the tube (Figures 2 and 3). Some extended worms appeared to be shy and quickly retracted into the tube when their pictures were taken (Figure 3). On some occasions, the worms showed high densities, either inside a living host (Figures 2A,B,D, 4E and 6A) or on dead coral (Figure 3C,D).

Some host coral species showed peak-shaped deformities around the worm tubes (Figure 4). In the foliaceous coral *Agaricia lamarcki*, the deformity resembles a sleeve that continues to grow upward and in thickness around the worm's tube, allowing the top to remain free (Figure 4A,B). Peak-shaped deformities in various sizes were most abundantly found in *Pseudodiploria strigosa* (Figure 4C–E) and less commonly in *Orbicella annularis* and *O. franksi* (Figure 4F,G). When the largest peak found in *P. strigosa* (Figure 4C) was removed, the worm tube appeared to be at least 8 cm long and deeply embedded inside the remaining part of the host coral (Figure 5).

Coral injuries were abundant around worm tubes in various coral species (Figures 6 and 7). The wounds, visible as dead lesions, were either at the periphery of live coral tissue (Figure 6A,B) or more toward the middle and surrounded by live coral tissue (Figures 6C,D and 7). Some dead patches were used as substrates by algae and sponges (Figures 6 and 7). In some coral species, the live tissue around the gash showed a discoloration, suggesting that it was spreading from the wound centered around the worm (Figure 7C,D,F).

Table 1. Records of stony corals and other sessile invertebrates as host species (by family) for sabellid worms (*Anamobaea* sp.) based on photographs taken at Curaçao (a: 2021 and 2022; b: 2017; c: 2015; and d: 2014), Bonaire (e: 2019), and St. Eustatius (f: 2015).

Host Species	Curaçao	Bonaire	St. Eustatius
Cnidaria: Anthozoa: Scleractinia			
Agariciidae			
Agaricia agaricites (Linnaeus, 1758)	a	e	-
Agaricia fragilis Dana, 1846	a	-	f
Agaricia humilis (Verrill, 1901)	a	-	-
Agaricia lamarcki Milne Edwards & Haime, 1851	a,b,d	-	f
Helioseris cucullata (Ellis and Solander, 1786)	a	-	f
Astrocoeniidae			
Stephanocoenia intersepta (Esper, 1795)	a,b	-	f
Faviidae: Faviinae			
Colpophyllia natans (Houttuyn, 1772)	a,b	-	-
Diploria labyrinthiformis (Linnaeus, 1758)	a	-	f
Pseudodiploria strigosa (Dana, 1846)	a,d	-	-
Faviidae: Mussiinae			
Mycetophyllia aliciae Wells, 1973	-	-	f
Meandrinidae			
Eusmilia fastigiata (Pallas, 1766)	b	-	-
Meandrina jacksoni Weil & Pinzón, 2011	-	-	f
Meandrina meandrites (Linnaeus, 1758)	a,c	e	f
Merulinidae			
Dendrogyra cylindrus (Ehrenberg, 1834)	a	-	-
Orbicella annularis (Ellis & Solander, 1786)	a	-	f
Orbicella faveolata (Ellis & Solander, 1786)	a	e	-
Orbicella franksi (Gregory, 1895)	a,b,d	e	f
Montastraeidae			
Montastraea cavernosa (Linnaeus, 1767)	a,b	-	-
Pocilloporidae			
Madracis auretenra Locke, Weil & Coates, 2007	a,b	-	-
Madracis decactis (Lyman, 1859)	a	-	-
Madracis pharensis (Heller, 1868)	a,b	e	f
Madracis senaria Wells, 1973	a,b	e	-
Poritidae			
Porites astreoides Lamarck, 1816	a,d	e	f
Porites porites (Pallas, 1766)	-	-	f
Rhizangiidae			
Siderastrea siderea (Ellis & Solander, 1768)	a,b	-	f
Cnidaria: Hydrozoa			
Milleporidae			
Millepora alcicornis Linnaeus, 1758	a	e	-
Millepora complanata Lamarck, 1816	a	-	-
Cnidaria: Anthozoa: Alcyonacea			
Anthothelidae			
Erythropodium caribaeorum (Duchassaing & Michelotti, 1860)	a	-	-
Chordata: Tunicata: Ascidiacea			
Didemnidae			
Trididemnum solidum (Van Name, 1902)	a,b	-	-
Porifera spp.	a	-	-
Unidentified dead coral with/without algae	a	-	-

Figure 2. Split-crown feather dusters (*Anamobaea* sp.) hosted by scleractinian corals in the Dutch Caribbean. (**A**) *Diploria labyrinthiformis* at St. Eustatius (2015) hosting five extended worms (one next to the coral colony) and three contracted ones (arrows). (**B**) *Siderastrea siderea* at St. Eustatius (2015) with four extended worms (two next to the coral colony). (**C**) *Porites astreoides* at St. Eustatius (2015) showing two extended worms. (**D**) *Meandrina jacksoni* at St. Eustatius (2015) hosting seven extended worms. (**E**) *Madracis decactis* at Bonaire (2019) with two extended worms. (**F**) *Helioseris cucullata* at St. Eustatius (2015) with one extended worm.

Figure 3. Split-crown feather dusters (*Anamobaea* sp.) at Curaçao (2021). (**A**) A single worm on dead coral in the extended condition, showing its radioles. (**B**) The same worm withdrawn inside its tube, overgrown by filamentous algae. (**C**) Four worms on dead coral, one extended. (**D**) The same worms, all withdrawn. (**E**) Two extended worms in association with a *Millepora alcicornis* coral. (**F**) Both worms retracted. Arrows indicate worms that had just retracted. The maximum width of the worm tubes is ca. 5 mm.

Figure 4. Coral deformations caused by the presence of split-crown feather dusters (*Anamobaea* sp.) in various host coral species. (**A,B**) The host coral *Agaricia lamarcki* at Curaçao (2021) with two peaks in their initial phase (**A**: arrows) and a large peak (**B**: arrow). (**C**) Close-up of the coral *Pseudodiploria strigosa* at Curaçao (2021) showing a large peak. (**D,E**) Corals of *P. strigosa* at St. Eustatius (2015), one showing a peak with an extended worm inside (**D**: arrow) and another one with five worm peaks (**E**: arrows). (**F**) *Orbicella annularis* at Curaçao (2021) with one worm peak (arrow). (**G**) *Orbicella franksi* at St. Eustatius (2015) with a small worm peak (arrow), next to a serpulid Christmas tree worm (*S. giganteus*). The maximum width of each sabellid tube is ca. 5 mm.

Figure 5. Tube of a split-crown feather duster (*Anamobaea* sp.) after removal of the peak-shaped deformation in a *Pseudodiploria strigosa* coral (see Figure 4C). The visible part of the tube is 8 cm long.

Feather duster worms of two other species were not observed inside living corals but in dead skeleton directly next to a living coral or in a patch of dead coral surrounded by healthy coral tissue. They are the magnificent feather duster *Sabellastarte magnifica* (Shaw, 1800) (Figure 10) and the social feather duster *Bispira brunnea* (Treadwell, 1917) (Figure 11). *Sabellastarte magnifica* was found in or next to live coral colonies of the corals *Diploria labyrinthiformis*, *Madracis auretenra*, *Meandrina meandrites*, *Millepora alcicornis*, *Orbicella annularis*, *Pseudodiploria strigosa*, and *Stephanocoenia intersepta*. Their tubes reached diameters of nearly 2 cm and could therefore be distinguished from the tubes of *Anamobaea* sp., which reached up to 0.5 cm in width. *Bispira brunnea* was only found on dead patches of *Montastraea cavernosa* and *Orbicella annularis* (Figure 11). A published photograph from the Mexican Caribbean shows *B. brunnea* in a coral injury on top of a colony of *Siderastrea siderea* [54]. This worm species can be distinguished from the other two because it occurs as colonies instead of single individuals and because its tubes and radioles are much smaller than those of the others. Because all *Bispira* and *Sabellastarte* worms appeared to live on dead coral skeleton, near live coral, or at a distance, it is unclear whether they were symbionts or part of pseudo-associations.

4. Discussion

This report presents, for the first time, extensive evidence for the association of feather duster worms with corals, other sessile invertebrates, and algae in the Caribbean. This discovery is remarkable because of the strikingly large wounds and deformities inflicted by them on their host corals. Photographs of the worms indicate that these associations have been present at least since 2014 on the coral reefs of Curaçao (Southern Caribbean) and since 2015 at St. Eustatius (Eastern Caribbean). Prior to that, they may have remained unnoticed because of the worm's withdrawal behavior, because it was perhaps less abundant in the past, or because scientists studying the worms did not pay much attention to the hosts.

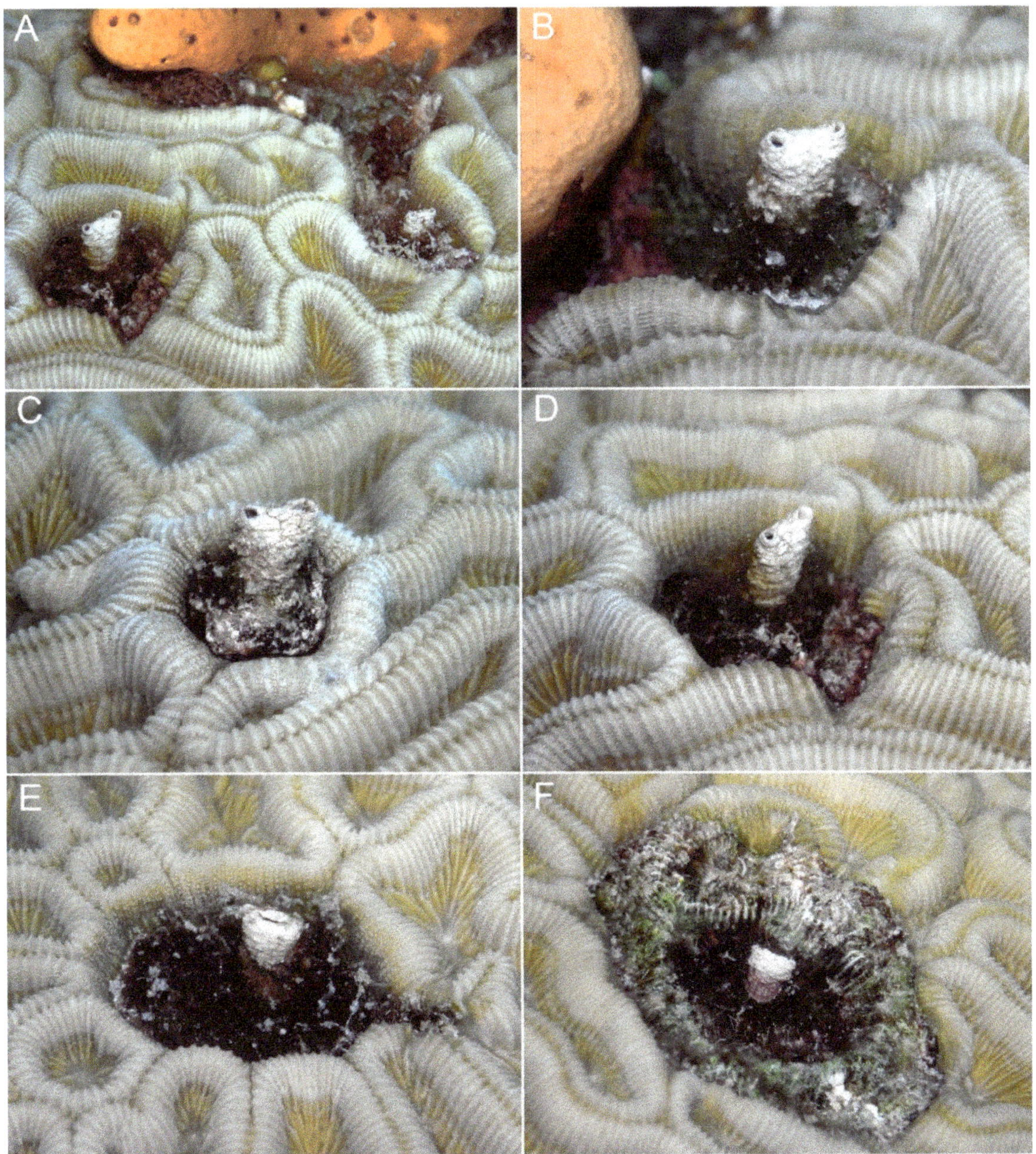

Figure 6. Overview (**A**) and close-up images (**B–F**) of coral damage caused by split-crown feather dusters (*Anamobaea* sp.) on a large *Colpophyllia natans* colony at Curaçao (2021). The images show various developmental stages of coral injuries (dead skeleton covered by turf algae) forming coves at the coral margin (**A,B**) and circular patches over the colony surface (**C–F**). The maximum width of each tube is ca. 5 mm.

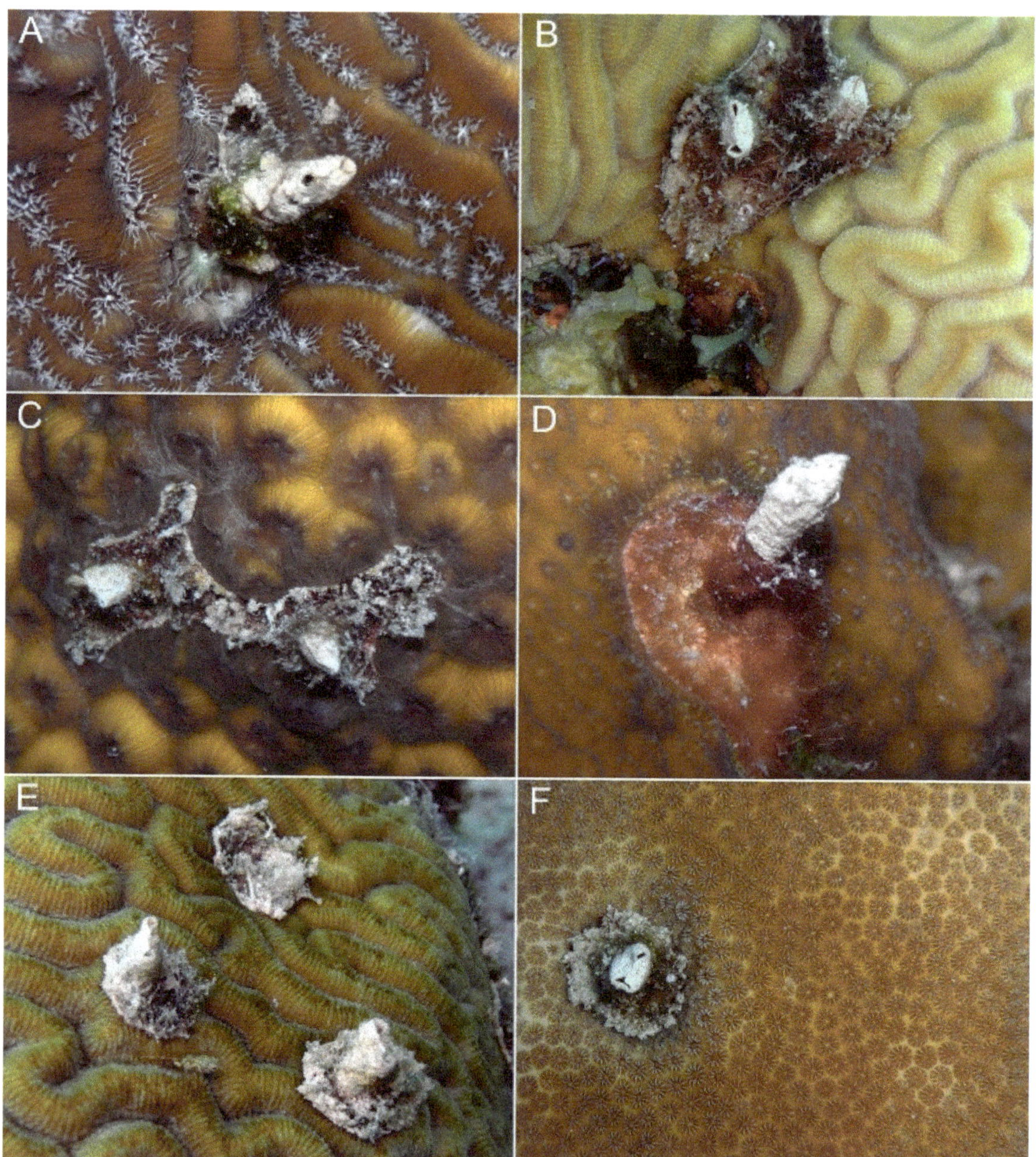

Figure 7. Close-up images of coral injuries around tubes made by split-crown feather dusters (*Anamobaea* sp.) shown in retracted condition at Curaçao (2021). The coral injuries are observed in various host species, such as (**A**) *Agaricia lamarcki*, (**B**) *Diploria labyrinthiformis*, (**C**) *Montastraea cavernosa*, (**D**) *Orbicella annularis*, (**E**) *Pseudodiploria strigosa*, and (**F**) *Stephanocoenia intersepta*. In some species, the live tissue around the wound shows a discoloration (**C,D,F**). The maximum width of each tube is ca. 5 mm.

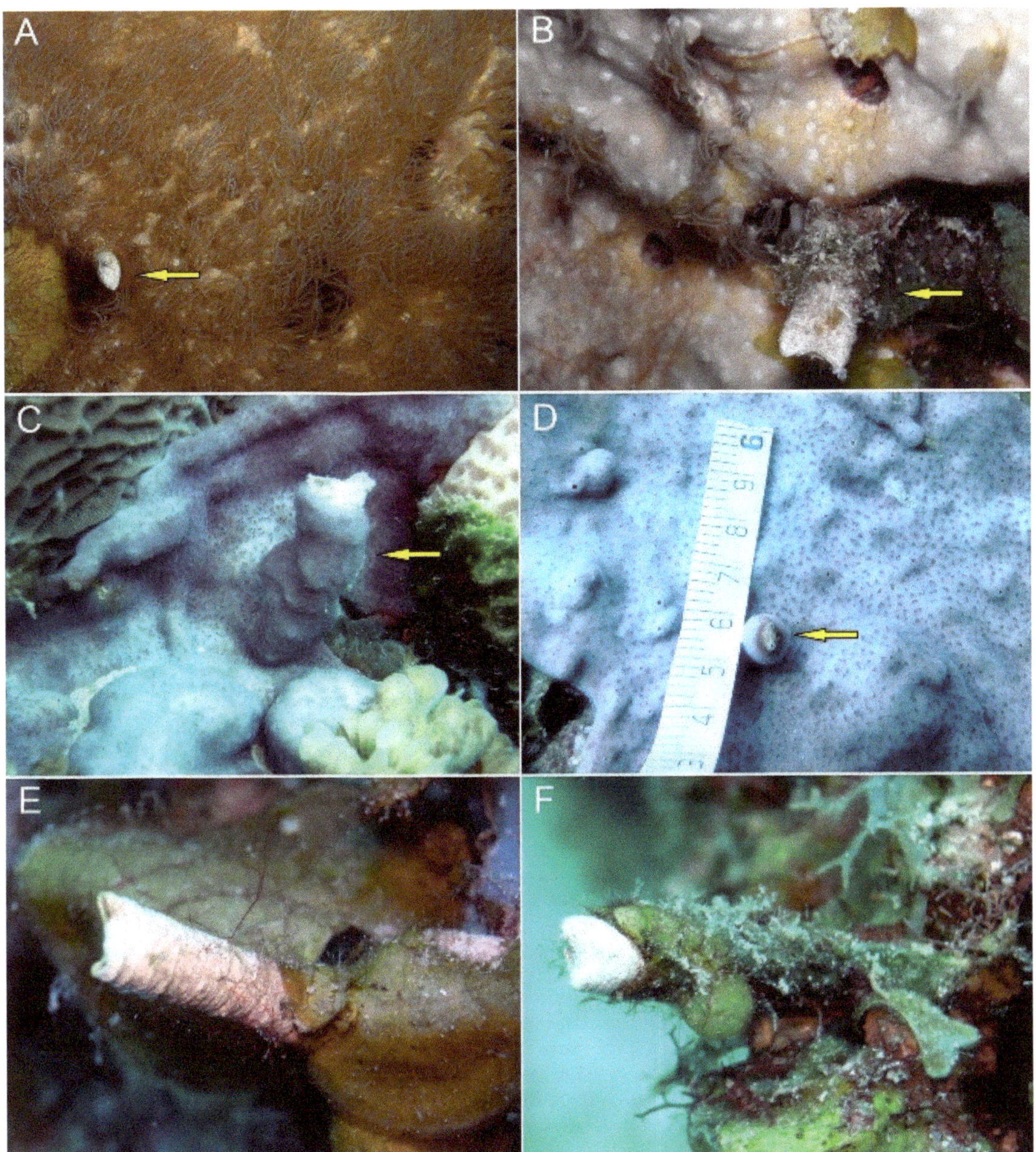

Figure 8. Split-crown feather dusters (*Anamobaea* sp.) hosted by noncoral invertebrates and algae that have overgrown corals: (**A**,**B**) The encrusting soft coral *Erythropodium caribaeorum* acting as a host on dead coral at Curaçao (2021), with tentacles extended (**A**) and retracted (**B**). (**C**,**D**) The encrusting colonial ascidian *Trididemnum solidum* at Curaçao overgrowing scleractinian host corals and worm tubes (except for the tube opening): on a scleractinian coral *Eusmilia fastigiata* in 2017 (**C**) and on dead coral in 2021 (**D**). (**E**,**F**) The phaeophyceaen alga *Lobophora* sp. at Curaçao (2021). Arrows indicate worm tubes. The maximum width of each tube is ca. 5 mm.

Figure 9. Split-crown feather dusters (*Anamobaea* sp.) at Curaçao (2021 and 2022) hosted by sponges that probably act as secondary hosts: (**A**) An unidentified black sponge partly overgrowing a worm tube and its host coral, *Siderastrea siderea*. (**B**) A zoantharian-infested sponge, *Niphates* sp., with an expanded worm. (**C**) A dark-red sponge, *Plakortis* sp., with a worm tube (arrow) next to the original host coral, *Orbicella franksi*. (**D**) An orange-red sponge, *Scopalina ruetzleri* (Wiedenmayer, 1977) with one worm tube (arrow). The maximum width of each tube is ca. 5 mm.

Coral deformities around sabellid worms embedded in the host's skeleton appear to be limited to a few scleractinian species of which *Pseudodiploria strigosa* appears to be the most common. Because coral-dwelling sabellids have been observed deep inside the coral skeleton, and the life span of sabellids may be over 10 years [55], these deformations have taken several years to develop. Morphological anomalies are not exceptional among corals inhabited by associated fauna. For example, the sabellid *Perkinsiana anodina* lives in short tube-shaped protuberances on the surface of an encrusting mushroom coral, which are part of the host's coral skeleton [32,51]. Coral gall crabs have received much attention because of the crescent-, canopy, slit-, and basket-shaped pits inside various coral species [44,56–59]. Coral cysts and pits made by other crabs and by shrimps in stony corals have also been described [60,61], which should not be confused with gall-shaped excavations made by shrimps [62,63], bivalves [64–67], and gastropods [34,68]. Copepods are also known to induce the forming of dwellings in corals, either as galls [69,70] or as tubular outgrowths [71,72]. Ascothoracidan crustaceans of the genus *Petrarca* Fowler, 1889 are known to form conspicuous galls in shallow-water and deep-sea corals [73–75].

Figure 10. Magnificent feather dusters (*Sabellastarte magnifica*) at Curaçao (2022) in close proximity to corals. (**A**) An extended worm on a dead coral patch of *Orbicella annularis*. (**B**) Same individual retracted inside its tube. (**C**) A worm in a colony of *Madracis auretenra* surrounded by healthy branch tips but attached to their dead base. (**D**) A worm surrounded by colonies of *Millepora alcicornis* and *Pseudodiploria strigosa*. (**E**) A worm underneath a colony of *Diploria labyrinthiformis*. (**F**) A worm on a dead patch of *Orbicella annularis* surrounded by healthy coral tissue. The width of each worm tube is ca. 2 cm.

Figure 11. Social feather dusters (*Sabellastarte magnifa*) at Curaçao (2022) in close proximity to live coral. (**A**) A worm colony on a dead patch of *Orbicella annularis*. (**B**) Another worm colony on dead coral underneath a *Montastraea cavernosa*.

Tube-dwelling gammarid amphipods and chaetopterid polychaete worms have been reported to induce the forming of densely distributed finger-like structures in *Montipora* corals [76,77]. Coral barnacles usually become embedded in the coral skeleton and become partly overgrown by coral tissue [44,60]. Some alterations in the coral skeleton morphology are microscopic and hardly visible, such as those caused by coral-dwelling hydroids of the genus *Zanclea* [78–80]. In contrast, vermetid snails that live inside branching *Stylophora* and massive *Porites* corals are known to modify the host's morphology on a larger scale by flattening its surface relief, which is attributed to growth inhibition caused by the snail's toxic mucus webs [81–83]. In contrast, large growth alterations in massive, branching, and encrusting corals consisting of deep fissures can be formed by *Pedum* scallops embedded in corals [34,66,67]. Some aggressive coral-dwelling sponges are not considered long-term associated fauna because they usually tend to overgrow and kill their hosts, but in some foliose corals, they evoke a morphological response, which is visible as the growth of flap-like protrusions that overlap the approaching sponges [84]. A modified morphology is also seen in other foliose corals that overgrow sponges as if the coral shape is molded by that of the sponge [85]. All these examples indicate that some corals may adapt their shape to resist the presence of potentially harmful associated fauna or competitors for space.

Coral injuries caused by feather duster worms have not been reported before. These appear to be much larger than those caused by coral-dwelling *Spirobranchus* worms [18,19,86]. On the other hand, large densities of serpulid worms overgrowing live coral may eventually cause partial coral mortality [87]. Many feather duster worms in the present study were found on dead coral (Figure 3A,B), and some of them formed clusters (Figure 3C,D). In some cases, the worm-infested dead-coral area was next to live coral, suggesting that the worms contributed to partial coral mortality (Figures 3E,F and 5A). A few patches of dead coral were surrounded by discolored live-coral tissue (Figure 6C,D,F). This may represent a reaction to stress as seen in some massive *Porites* corals in which polyps in contact with algae or epifauna show pink or purple pigmentation [18,66,88–90]. The difference is that there may not be extra pigmentation in the examples of the present study.

Some split-crown feather dusters at Curaçao were not hosted by stony corals but by other invertebrates. These invertebrates may have either colonized dead coral or overgrown living corals and became secondary hosts when the worms were able to resist becoming overgrown as well. The last scenario has been shown by serpulid Christmas tree worms of the genus *Spirobranchus*. The encrusting octocoral *Erythropodium caribaeorum* is recognized as an aggressive competitor for space in the Caribbean [91], which is able to overgrow corals but apparently not their symbiotic *Spirobranchus* worms [92], similar to some feather duster worms at Curaçao in the present study (Figure 8A,B). Similarly, the colonial tunicate *Trididemnum solidum* is notorious for overgrowing Caribbean corals [93,94], except for their associated *Spirobranchus* [95] and seemingly also individuals of symbiotic *Anamobaea* sp. (Figure 8C,D). Sponges are also able to overgrow corals with the exception of symbiotic *Spirobranchus* [96,97] and apparently also *Anamobaea* sp. (Figure 9). The feather duster worm was also observed in association with algae, in particular the brown algae *Lobophora* sp. (Figure 8E,F). *Lobophora* has increased in abundance over the last decades at Curaçao and is able to overgrow live coral [98,99]. It is likely that it is able to overgrow dead and live coral containing *Anamobaea* sp., but apparently the worm tubes protrude too far to become outcompeted.

The cause of the injurious effect of the feather duster worms is unclear. The size of the wounds suggests that the worms produce toxins, but there is limited information on toxicity as a defense mechanism in Sabellidae [100]. The use of toxins can perhaps prevent worms from becoming overgrown by their hosts, as seen in *Pedum* scallops [66]. The mucus secreted by some sabellid species proves to have antibacterial properties [101–103]. According to a recent review paper on polychaete toxins, no relevant information appears to be available on the negative effect of sabellid mucus on other organisms [104]. In contrast, coral-dwelling worm snails, which occupy the same ecological niche as the feather duster worms of the present study [20], are well known for their venomous mucus and the damage this may inflict on the host corals [105,106].

Unlike *Anamobaea* sp., the relation of *Sabellastarte magnifica* and *Bispira brunnea* to corals is unclear because they were never found in living coral tissue (Figures 10 and 11). A close proximity to live corals shown by these two species may be unusual since they were also commonly found at a distance from live corals. Therefore, it may be more appropriate to use the term "pseudo-association" for this kind of unclear relation. On the other hand, it is also possible that these worms cause damage to corals and are responsible for coral mortality in their proximity.

The present study shows that the Caribbean feather duster worm *Anamobaea* sp. is more common and harmful to corals than previously known. The species has a symbiotic relation with a large range of corals and other invertebrates, which was also unknown before. It is unclear if the species has increased in abundance recently. Because this worm has the potential to become a pest species, future research should focus on its population dynamics, its settling behavior on live corals (as done with larvae of symbiotic barnacles [107,108]), and the cause, growth, and extension of coral wounds around its tubes. The larval settlement behavior of *S. magnifica* and *B. brunnea* also needs to be investigated in order to find out whether these species prefer to live in close proximity to corals or not.

Author Contributions: Conceptualization, B.W.H. and R.J.v.d.S.; methodology, B.W.H., C.E.H., S.J.L.-D., R.J.v.d.S., A.S.-M., R.S., and R.F.T.; validation, B.W.H.; formal analysis, B.W.H.; investigation, B.W.H., C.E.H., S.J.L.-D., R.J.v.d.S., A.S.-M., R.S., and R.F.T.; resources, B.W.H., C.E.H., S.J.L.-D., R.J.v.d.S., A.S.-M., R.S., and R.F.T.; data curation, B.W.H., C.E.H., S.J.L.-D, R.J.v.d.S., A.S.-M., R.S., and R.F.T.; writing—original draft preparation, B.W.H.; writing—review and editing, B.W.H., C.E.H., S.J.L.-D., R.J.v.d.S., A.S.-M., R.S., and R.F.T.; visualization, B.W.H., C.E.H., S.J.L.-D., R.J.v.d.S., A.S.-M., R.S., and R.F.T.; supervision, B.W.H.; project administration, B.W.H.; funding acquisition, B.W.H., C.E.H., S.J.L.-D., R.J.v.d.S., A.S.-M., R.S., and R.F.T. All authors have read and agreed to the published version of the manuscript.

Funding: The field research at Curaçao was funded by the Alida M. Buitendijk Fund, the Jan-Joost ter Pelkwijk Fund, the Holthuis Fund, the Groningen University Fund, and the Dutch Research Council (NWO) Doctoral Grant for Teachers Programme (nr. 023.015.036). Fieldwork at Bonaire was supported by the World Wildlife Fund (WWF) Netherlands. The Treub Maatschappij (Society for the Advancement of Research in the Tropics) funded research at Bonaire and Curaçao.

Institutional Review Board Statement: Not applicable.

Informed Consent Statement: Not applicable.

Data Availability Statement: Data sharing not applicable.

Diversity **2022**, 14, 332

Acknowledgments: We are grateful to the funding agencies mentioned above. We thank María Ana Tovar-Hernández (Universidad Autónoma de Nuevo León, Mexico) for confirming the identity of the sabellid worms and Jaaziel E. García-Hernández for confirming the identity of the host sponges. We thank the staff of CARMABI (Curaçao) and the Dive Shop for their hospitality and assistance during the fieldwork. BWH is also grateful to Stichting Nationale Parken Bonaire (STINAPA), the Dutch Caribbean Nature Alliance (DCNA) and Dive Friends (Bonaire) for logistic support at Bonaire, the Caribbean Netherlands Science Institute (CNSI), St. Eustatius National Parks Foundation (STENAPA), and Scubaqua Dive Center for facilitating research at St. Eustatius. We want to thank three anonymous reviewers for their constructive comments, which helped us to improve the manuscript.

Conflicts of Interest: The authors declare no conflict of interest.

References

1. Reaka-Kudla, M.L. The global biodiversity of coral reefs: A comparison with rain forests. In *Biodiversity II: Understanding and Protecting Our Natural Resources*; Reaka-Kudla, M.L., Wilson, D.E., Wilson, E.O., Eds.; Joseph Henry/National Academy Press: Washington, DC, USA, 1997; pp. 83–108.
2. Knowlton, N.; Brainard, R.E.; Fisher, R.; Moews, M.; Plaisance, L.; Caley, M.J. Coral reef biodiversity. In *Life in the World's Oceans: Diversity, Distribution and Abundance*; McIntyre, A.D., Ed.; Wiley-Blackwell: Chichester, UK, 2010; pp. 65–74.
3. Glynn, P.W.; Enochs, I.C. Invertebrates and their roles in coral reef ecosystems. In *Coral Reefs: An Ecosystem in Transition*; Dubinsky, Z., Stambler, N., Eds.; Springer: Dordrecht, The Netherlands, 2011; pp. 273–325. [CrossRef]
4. Nelson, H.R.; Kuempel, C.D.; Altieri, A.H. The resilience of reef invertebrate biodiversity to coral mortality. *Ecosphere* **2016**, 7, e01399. [CrossRef]
5. Castro, P. Animal symbioses in coral reef communities: A review. *Symbiosis* **1988**, 5, 161–184.
6. Montano, S. The extraordinary importance of coral-associated fauna. *Diversity* **2020**, 12, 357. [CrossRef]
7. Scaps, P.; Denis, V. Can organisms associated with live scleractinian corals be used as indicators of coral reef status? *Atoll Res. Bull.* **2008**, 566, 1–18. [CrossRef]
8. Stella, J.S.; Pratchett, M.S.; Hutchings, P.A.; Jones, G.P. Coral-associated invertebrates: Diversity, ecology importance and vulnerability to disturbance. *Oceanogr. Mar. Biol. Ann. Rev.* **2011**, 49, 43–104.
9. DeVantier, L.M.; Reichelt, R.E.; Bradbury, R.H. Does *Spirobranchus giganteus* protect host *Porites* from predation by *Acanthaster planci*: Predator pressure as a mechanism of coevolution? *Mar. Ecol. Prog. Ser.* **1986**, 32, 307–310. [CrossRef]
10. Pratchett, M.S. Influence of coral symbionts on feeding preferences of crown-of-thorns starfish *Acanthaster planci* in the western Pacific. *Mar. Ecol. Prog. Ser.* **2001**, 214, 111–119. [CrossRef]
11. Ben-Tzvi, O.; Einbinder, S.; Brokovich, E. A beneficial association between a polychaete worm and a scleractinian coral? *Coral Reefs* **2006**, 25, 98. [CrossRef]
12. Rouzé, H.; Lecellier, G.; Mills, S.C.; Planes, S.; Berteaux-Lecellier, V.; Stewart, H. Juvenile *Trapezia* spp. crabs can increase juvenile host coral survival by protection from predation. *Mar. Ecol. Prog. Ser.* **2014**, 515, 151–159. [CrossRef]
13. Montano, S.; Fattorini, S.; Parravicini, V.; Berumen, M.L.; Galli, P.; Maggioni, D.; Arrigoni, R.; Seveso, D.; Strona, G. Corals hosting symbiotic hydrozoans are less susceptible to predation and disease. *Proc. R. Soc. B* **2017**, 284, 20172405. [CrossRef]
14. Stewart, H.L.; Holbrook, S.J.; Schmitt, R.J.; Brooks, A.J. Symbiotic crabs maintain coral health by clearing sediments. *Coral Reefs* **2006**, 25, 609–615. [CrossRef]
15. Shima, J.S.; Osenberg, C.W.; Stier, A. The vermetid gastropod *Dendropoma maximum* reduces coral growth and survival. *Biol. Lett.* **2010**, 6, 815–818. [CrossRef] [PubMed]
16. Carballo, J.L.; Bautista, E.; Nava, H.; Cruz-Barraza, J.A.; Chávez, J.A. Boring sponges, an increasing threat for coral reefs affected by bleaching events. *Ecol. Evol.* **2013**, 3, 872–886. [CrossRef] [PubMed]
17. De Bakker, D.M.; Webb, A.E.; van den Bogaart, L.A.; van Heuven, S.M.A.C.; Meesters, E.H.; van Duyl, F.C. Quantification of chemical and mechanical bioerosion rates of six Caribbean excavating sponge species found on the coral reefs of Curaçao. *PLoS ONE* **2018**, 13, e0197824. [CrossRef] [PubMed]
18. Hoeksema, B.W.; van der Schoot, R.J.; Wels, D.; Scott, C.M.; ten Hove, H.A. Filamentous turf algae on tube worms intensify damage in massive *Porites* corals. *Ecology* **2019**, 100, e02668. [CrossRef] [PubMed]
19. Hoeksema, B.W.; Wels, D.; van der Schoot, R.J.; ten Hove, H.A. Coral injuries caused by *Spirobranchus* opercula with and without epibiotic turf algae at Curaçao. *Mar. Biol.* **2019**, 166, 60. [CrossRef]
20. Hoeksema, B.W.; Harper, C.E.; Langdon-Down, S.J.; van der Schoot, R.J.; Smith-Moorhouse, A.; Spaargaren, R.; Timmerman, R.F. Host range of the coral-associated worm snail *Petaloconchus* sp. (Gastropoda: Vermetidae), a newly discovered cryptogenic pest species in the southern Caribbean. *Diversity* **2022**, 14, 196. [CrossRef]
21. Vinn, O.; Zatoń, M.; Tovar-Hernández, M.A. Tube microstructure and formation in some feather duster worms (Polychaeta, Sabellidae). *Mar. Biol.* **2018**, 165, 98. [CrossRef]
22. Humann, P.; Deloach, N.; Wilk, L. *Reef Creature Identification: Florida, Caribbean, Bahamas*, 3rd ed.; New World Publications: Jacksonville, FL, USA, 2013; 366p.

23. Bok, M.J.; Capa, M.; Nilsson, D.E. Here, there and everywhere: The radiolar eyes of fan worms (Annelida, Sabellidae). *Integr. Comp. Biol.* **2016**, *56*, 784–795. [CrossRef]

24. Capa, M.; Kupriyanova, E.; Nogueira, J.M.D.M.; Bick, A.; Tovar-Hernández, M.A. Fanworms: Yesterday, today and tomorrow. *Diversity* **2021**, *13*, 130. [CrossRef]

25. Capa, M.; López, E. Sabellidae (Annelida: Polychaeta) living in blocks of dead coral in the Coiba National Park, Panamá. *J. Mar. Biol. Assoc. U. K.* **2004**, *84*, 63–72. [CrossRef]

26. Tovar-Hernández, M.A.; Knight-Jones, P. Species of *Branchiomma* (Polychaeta: Sabellidae) from the Caribbean Sea and Pacific coast of Panama. *Zootaxa* **2006**, *1189*, 1–37. [CrossRef]

27. Tovar-Hernández, M.A.; Salazar-Vallejo, S.I. Sabellids (Polychaeta: Sabellidae) from the Grand Caribbean. *Zool. Stud.* **2006**, *45*, 24–66.

28. Giangrande, A.; Licciano, M.; Gambi, M.C. A collection of Sabellidae (Polychaeta) from Carrie Bow Cay (Belize, western Caribbean Sea) with the description of two new species. *Zootaxa* **2007**, *1650*, 41–53. [CrossRef]

29. Tovar-Hernández, M.A.; Salazar-Silva, P. Catalogue of Sabellidae (Annelida: Polychaeta) from the Grand Caribbean Region. *Zootaxa* **2008**, *1894*, 1–22. [CrossRef]

30. Bastida-Zavala, J.R.; Buelna, A.S.R.; De Leon-Gonzalez, J.A.; Camacho-Cruz, K.A.; Carmona, I. New records of sabellids and serpulids (Polychaeta: Sabellidae, Serpulidae) from the Tropical Eastern Pacific. *Zootaxa* **2016**, *4184*, 401–457. [CrossRef]

31. Dávila-Jiménez, Y.; Tovar-Hernández, M.A.; Simões, N. The social feather duster worm *Bispira brunnea* (Polychaeta: Sabellidae): Aggregations, morphology and reproduction. *Mar. Biol. Res.* **2017**, *13*, 782–796. [CrossRef]

32. Tovar-Hernández, M.A.; García-Garza, M.E.; de León-González, J.A. Sclerozoan and fouling sabellid worms (Annelida: Sabellidae) from Mexico with the establishment of two new species. *Biodivers. Data J.* **2020**, *8*, e57471. [CrossRef]

33. Patton, W.K. Animal associates of living reef corals. In *Biology and Geology of Coral Reefs III. Biology 2*; Jones, O.A., Endean, R., Eds.; Academic Press: New York, NY, USA, 1976; pp. 1–37.

34. Zann, L.P. *Living Together in the Sea*; T.F.H. Publications: Neptune, NY, USA, 1980; 416p.

35. Scott, P.J.B. Aspects of living coral associates in Jamaica. In Proceedings of the 5th International Coral Reef Congress, Tahiti, French Polynesia, 27 May–1 June 1985; Volume 5, pp. 345–350.

36. Scott, P.J.B. Associations between corals and macro-infaunal invertebrates in Jamaica, with a list of Caribbean and Atlantic coral associates. *Bull. Mar. Sci.* **1987**, *40*, 271–286.

37. Perry, C.T. Macroborers within coral framework at Discovery Bay, north Jamaica: Species distribution and abundance, and effects on coral preservation. *Coral Reefs* **1998**, *17*, 277–287. [CrossRef]

38. Lewis, J.B. Biology and ecology of the hydrocoral *Millepora* on coral reefs. *Adv. Mar. Biol.* **2000**, *50*, 1–55. [CrossRef]

39. Nogueira, J.M.M. Fauna living in colonies of *Mussismilia hispida* (Verrill) (Cnidaria: Scleractinia) in four South-eastern Brazil Islands. *Braz. Arch. Biol. Technol.* **2003**, *46*, 421–432. [CrossRef]

40. Castro, C.; Monroy, M.; Solano, O.D. Estructura de la comunidad epifaunal asociada a colonias de vida libre del hydrocoral *Millepora alcicornis* Linnaeus, 1758 en Bahía Portete, Caribe Colombiano. *Bol. Investig. Mar. Cost.* **2006**, *35*, 195–206.

41. Oigman-Pszczol, S.S.; Creed, J.C. Distribution and abundance of fauna on living tissues of two Brazilian hermatypic corals (*Mussismilia hispida* (Verril 1902) and *Siderastrea stellata* Verril, 1868). *Hydrobiologia* **2006**, *563*, 143–154. [CrossRef]

42. Stella, J.S.; Jones, G.P.; Pratchett, M.S. Variation in the structure of epifaunal invertebrate assemblages among coral hosts. *Coral Reefs* **2010**, *29*, 957–973. [CrossRef]

43. Hoeksema, B.W.; van der Meij, S.E.T.; Fransen, C.H.J.M. The mushroom coral as a habitat. *J. Mar. Biol. Assoc. UK* **2012**, *92*, 647–663. [CrossRef]

44. Hoeksema, B.W.; van Beusekom, M.; ten Hove, H.A.; Ivanenko, V.N.; van der Meij, S.E.T.; van Moorsel, G.W.N.M. *Helioseris cucullata* as a host coral at St. Eustatius. Dutch Caribbean. *Mar. Biodivers.* **2017**, *47*, 71–78. [CrossRef]

45. Lymperaki, M.M.; Hill, C.E.; Hoeksema, B.W. The effects of wave exposure and host cover on coral-associated fauna of a centuries-old artificial reef in the Caribbean. *Ecol. Eng.* **2022**, *176*, 106536. [CrossRef]

46. Martin, D.; Britayev, T.A. Symbiotic polychaetes: Review of known species. *Oceanogr. Mar. Biol. Ann. Rev.* **1998**, *36*, 217–340.

47. Molodtsova, T.N.; Britayev, T.A.; Martin, D. Cnidarians and their polychaete symbionts. In *The Cnidaria, Past, Present and Future*; Goffredo, S., Dubinsky, Z., Eds.; Springer: Cham, Switzerland, 2016; pp. 387–413. [CrossRef]

48. Martin, D.; Britayev, T.A. Symbiotic polychaetes revisited: An update of the known species and relationships (1998–2017). *Oceanogr. Mar. Biol. Ann. Rev.* **2018**, *56*, 371–447. [CrossRef]

49. Nogueira, J.M.M.; Knight-Jones, P. A new species of *Pseudobranchiomma* Jones (Polychaeta: Sabellidae) found amongst Brazilian coral, with a redescription of *P. punctata* (Treadwell, 1906) from Hawaii. *J. Nat. Hist.* **2002**, *36*, 1661–1670. [CrossRef]

50. Nogueira, J.M.M.; Amaral, A.C.Z. *Amphicorina schlenzae*, a small sabellid (Polychaeta, Sabellidae) associated with a stony coral on the coast of Sao Paulo State, Brazil. *Bull. Mar. Sci.* **2000**, *67*, 617–624.

51. Tovar-Hernández, M.A.; ten Hove, H.A.; Vinn, O.; Zatoń, M.; de León-González, J.A.; García-Garza, M.E. Fan worms (Annelida: Sabellidae) from Indonesia collected by the Snellius II Expedition (1984) with descriptions of three new species and tube microstructure. *PeerJ* **2020**, *8*, e9692. [CrossRef]

52. Benzoni, F.; Arrigoni, R.; Stefani, F.; Reijnen, B.T.; Montano, S.; Hoeksema, B.W. Phylogenetic position and taxonomy of *Cycloseris explanulata* and *C. wellsi* (Scleractinia: Fungiidae): Lost mushroom corals find their way home. *Contrib. Zool.* **2012**, *81*, 125–146. [CrossRef]

53. Nishi, E.; Gil, J.; Tanaka, K.; Kupriyanova, E.K. *Notaulax yamasui* sp. n. (Annelida, Sabellidae) from Okinawa and Ogasawara, Japan, with notes on its ecology. *ZooKeys* **2017**, *660*, 1–16. [CrossRef]

54. Tovar-Hernández, M.A.; Pineda-Vera, A. Taxonomía y estrategias reproductivas del poliqueto sabélido *Bispira brunnea* (Treadwell, 1917) del Caribe Mexicano. *Cienc. Mar.* **2008**, *21*, 3–14.

55. Nishi, E.; Nishihira, M. Use of annual density banding to estimate longevity of infauna of massive corals. *Fish. Sci.* **1999**, *65*, 48–56. [CrossRef]

56. van der Meij, S.E.T.; Fransen, C.H.J.M.; Pasman, L.R.; Hoeksema, B.W. Phylogenetic ecology of gall crabs (Cryptochiridae) as associates of mushroom corals (Fungiidae). *Ecol. Evol.* **2015**, *5*, 5770–5780. [CrossRef]

57. Klompmaker, A.A.; Portell, R.W.; van der Meij, S.E.T. Trace fossil evidence of coral-inhabiting crabs (Cryptochiridae) and its implications for growth and paleobiogeography. *Sci. Rep.* **2016**, *6*, 23443. [CrossRef]

58. Terrana, L.; Caulier, G.; Todinanahary, G.; Lepoint, G.; Eeckhout, I. Characteristics of the infestation of *Seriatopora* corals by the coral gall crab *Hapalocarcinus marsupialis* Stimpson, 1859 on the great reef of Toliara, Madagascar. *Symbiosis* **2016**, *69*, 113–122. [CrossRef]

59. Hoeksema, B.W.; Butôt, R.; García-Hernández, J.E. A new host and range record for the gall crab *Fungicola fagei* as a symbiont of the mushroom coral *Lobactis scutaria* in Hawai'i. *Pac. Sci.* **2018**, *72*, 251–261. [CrossRef]

60. Chan, B.K.K.; Wong, K.J.H.; Cheng, Y.R. Biogeography and host usage of coral-associated crustaceans: Barnacles, copepods, and gall crabs as model organisms. In *The Natural History of the Crustacea: Evolution and Biogeography of the Crustacea*; Thiel, M., Poore, G.C., Eds.; Oxford University Press: Oxford, UK, 2020; Volume 8, pp. 183–215. [CrossRef]

61. Hoeksema, B.W.; García-Hernández, J.E. Host-related morphological variation of dwellings inhabited by the crab *Domecia acanthophora* in the corals *Acropora palmata* and *Millepora complanata* (Southern Caribbean). *Diversity* **2020**, *12*, 143. [CrossRef]

62. Bruce, A.J. Notes on some Indo-Pacific Pontoniinae. XIV. Observations on *Paratypton siebenrocki* Balss. *Crustaceana* **1969**, *17*, 171–186. [CrossRef]

63. Scott, P.J.B.; Reiswigh, M.; Marcottbe, M. Ecology, functional morphology, behaviour, and feeding in coral- and sponge-boring species of *Upogebia* (Crustacea: Decapoda: Thalassinidea). *Can. J. Zool.* **1988**, *66*, 483–495. [CrossRef]

64. Scott, P.J.B. Distribution, habitat and morphology of the Caribbean coral-and rock-boring bivalve, *Lithophaga bisulcata* (d'Orbigny) (Mytilidae: Lithophaginae). *J. Molluscan Stud.* **1988**, *54*, 83–95. [CrossRef]

65. Kleemann, K. Boring and growth in chemically boring bivalves from the Caribbean, Eastern Pacific and Australia's Great Barrier Reef. *Senckenb. Marit.* **1990**, *22*, 101–154.

66. Chan, B.K.K.; Tan, J.C.H.; Ganmanee, M. Living in a growing host: Growth pattern and dwelling formation of the scallop *Pedum spondyloideum* in massive *Porites* spp. corals. *Mar. Biol.* **2020**, *167*, 95. [CrossRef]

67. Scaps, P. Association between the scallop *Pedum spondyloideum* (Bivalvia: Pteriomorphia: Pectinidae) and scleractinian corals from Nosy Be, Madagascar. *Cah. Biol. Mar.* **2020**, *61*, 73–80. [CrossRef]

68. Gittenberger, A.; Gittenberger, E. Cryptic, adaptive radiation of endoparasitic snails: Sibling species of *Leptoconchus* (Gastropoda: Coralliophilidae) in corals. *Org. Divers. Evol.* **2011**, *11*, 21–41. [CrossRef]

69. Dojiri, M. *Isomolgus desmotes*, new genus, new species (Lichomolgidae), a gallicolous poecilostome copepod from the scleractinian coral *Seriatopora hystrix* Dana in Indonesia, with a review of gall-inhabiting crustaceans of anthozoans. *J. Crust. Biol.* **1988**, *8*, 99–109. [CrossRef]

70. Kim, I.-H.; Yamashiro, H. Two species of poecilostomatoid copepods inhabiting galls on scleractinian corals in Okinawa, Japan. *J. Crust. Biol.* **2007**, *27*, 319–326. [CrossRef]

71. Ivanenko, V.N.; Moudrova, S.V.; Bouwmeester, J.; Berumen, M.L. First report of tubular corallites on *Stylophora* caused by a symbiotic copepod crustacean. *Coral Reefs* **2014**, *33*, 637. [CrossRef]

72. Shelyakin, P.V.; Garushyants, S.K.; Nikitin, M.A.; Mudrova, S.V.; Berumen, M.; Speksnijder, A.G.; Hoeksema, B.W.; Fontaneto, D.; Gelfand, M.S.; Ivanenko, V.N. Microbiomes of gall-inducing copepod crustaceans from the corals *Stylophora pistillata* (Scleractinia) and *Gorgonia ventalina* (Alcyonacea). *Sci. Rep.* **2018**, *8*, 11563. [CrossRef] [PubMed]

73. Grygier, M.J.; Cairns, S.D. Suspected neoplasms in deep-sea corals (Scleractinia: Oculinidae: *Madrepora* spp.) reinterpreted as galls caused by *Petrarca madreporae* n. sp. (Crustacea: Ascothoracida: Petrarcidae). *Dis. Aquat. Org.* **1996**, *24*, 61–69. [CrossRef]

74. Tachikawa, H.; Grygier, M.J.; Cairns, S.D. Live specimens of the parasite *Petrarca madreporae* (Crustacea: Ascothoracida) from the deep-water coral *Madrepora oculata* in Japan, with remarks on the development of its spectacular galls. *J. Mar. Sci. Technol.* **2020**, *28*, 58–64. [CrossRef]

75. Chan, B.K.K.; Dreyer, N.; Gale, A.S.; Glenner, H.; Ewers-Saucedo, C.; Pérez-Losada, M.; Kolbasov, G.A.; Crandall, K.A.; Høeg, J.T. The evolutionary diversity of barnacles, with an updated classification of fossil and living forms. *Zool. J. Linn. Soc.* **2021**, *193*, 789–846. [CrossRef]

76. Bergsma, G.S. Tube-dwelling coral symbionts induce significant morphological change in *Montipora*. *Symbiosis* **2009**, *49*, 143–150. [CrossRef]

77. Bergsma, G.S.; Martinez, C.M. Mutualist-induced morphological changes enhance growth and survival of corals. *Mar. Biol.* **2011**, *158*, 2267–2277. [CrossRef]

78. Manca, F.; Puce, S.; Caragnano, A.; Maggioni, D.; Pica, D.; Seveso, D.; Galli, P.; Montano, S. Symbiont footprints highlight the diversity of scleractinian-associated *Zanclea* hydrozoans (Cnidaria, Hydrozoa). *Zool. Scr.* **2019**, *48*, 399–410. [CrossRef]

79. Maggioni, D.; Arrigoni, R.; Seveso, D.; Galli, G.; Berumen, M.L.; Denis, V.; Hoeksema, B.W.; Huang, D.; Manca, F.; Pica, D.; et al. Evolution and biogeography of the *Zanclea*-Scleractinia symbiosis. *Coral Reefs* **2020**. [CrossRef]

80. Maggioni, D.; Saponari, L.; Seveso, D.; Galli, P.; Schiavo, A.; Ostrovsky, A.N.; Montano, S. Green fluorescence patterns in closely related symbiotic species of *Zanclea* (Hydrozoa, Capitata). *Diversity* **2020**, *12*, 78. [CrossRef]

81. Colgan, M.W. Growth rate reduction and modification of a coral colony by a vermetid mollusc *Dendropoma maxima*. In Proceedings of the 5th International Coral Reef Congress, Tahiti, French Polynesia, 27 May–1 June 1985; Volume 6, pp. 205–210.

82. Zvuloni, A.; Armoza-Zvuloni, R.; Loya, Y. Structural deformation of branching corals associated with the vermetid gastropod *Dendropoma maxima*. *Mar. Ecol. Prog. Ser.* **2008**, *363*, 103–108. [CrossRef]

83. Shima, J.S.; McNaughtan, D.; Strong, A.T. Vermetid gastropods mediate within-colony variation in coral growth to reduce rugosity. *Mar. Biol.* **2015**, *162*, 1523–1530. [CrossRef]

84. Elliott, J.; Patterson, M.; Vitry, E.; Summers, N.; Miternique, C. Morphological plasticity allows coral to actively overgrow the aggressive sponge *Terpios hoshinota* (Mauritius, Southwestern Indian Ocean). *Mar. Biodivers.* **2016**, *46*, 489–493. [CrossRef]

85. García-Hernández, J.E.; van Moorsel, G.W.N.M.; Hoeksema, B.W. Lettuce corals overgrowing tube sponges at St. Eustatius, Dutch Caribbean. *Mar. Biodivers.* **2017**, *47*, 55–56. [CrossRef]

86. Hoeksema, B.W.; ten Hove, H.A.; Berumen, M.L. A three-way association causing coral injuries in the Red Sea. *Bull. Mar. Sci.* **2018**, *94*, 1525–1526. [CrossRef]

87. Samimi Namin, K.; Risk, M.J.; Hoeksema, B.W.; Zohari, Z.; Rezai, H. Coral mortality and serpulid infestations associated with red tide, in the Persian Gulf. *Coral Reefs* **2010**, *29*, 509. [CrossRef]

88. Benzoni, F.; Galli, P.; Pichon, M. Pink spots on *Porites*: Not always a coral disease. *Coral Reefs* **2010**, *29*, 153. [CrossRef]

89. Benzoni, F.; Basso, D.; Caragnano, A.; Rodondi, G. *Hydrolithon* spp. (Rhodophyta, Corallinales) overgrow live corals (Cnidaria, Scleractinia) in Yemen. *Mar. Biol.* **2011**, *158*, 2419–2428. [CrossRef]

90. Kubomura, T.; Wee, H.B.; Reimer, J.D. Investigating incidence and possible causes of pink and purple pigmentation response in hard coral genus *Porites* around Okinawajima Island, Japan. *Reg. Stud. Mar. Sci.* **2021**, *41*, 101569. [CrossRef]

91. Ladd, M.C.; Shantz, A.A.; Burkepile, D.E. Newly dominant benthic invertebrates reshape competitive networks on contemporary Caribbean reefs. *Coral Reefs* **2019**, *38*, 1317–1328. [CrossRef]

92. Hoeksema, B.W.; Lau, Y.W.; ten Hove, H.A. Octocorals as secondary hosts for Christmas tree worms off Curaçao. *Bull. Mar. Sci.* **2015**, *91*, 489–490. [CrossRef]

93. Bak, R.P.M.; Sybesma, J.; van Duyl, F.C. The ecology of the tropical compound ascidian *Trididemnum solidum*. II. Abundance, growth and survival. *Mar. Ecol. Prog. Ser.* **1981**, *6*, 43–52. [CrossRef]

94. Sommer, B.; Harrison, P.L.; Scheffers, S.R. Aggressive colonial ascidian impacting deep coral reefs at Bonaire, Netherlands Antilles. *Coral Reefs* **2010**, *29*, 245. [CrossRef]

95. Hoeksema, B.W.; García-Hernández, J.E.; van Moorsel, G.W.N.M.; Olthof, G.; ten Hove, H.A. Extension of the recorded host range of Caribbean Christmas tree worms (*Spirobranchus* spp.) with two scleractinians, a zoantharian, and an ascidian. *Diversity* **2020**, *12*, 115. [CrossRef]

96. Hoeksema, B.W.; ten Hove, H.A.; Berumen, M.L. Christmas tree worms evade smothering by a coral-killing sponge in the Red Sea. *Mar. Biodivers.* **2016**, *48*, 15–16. [CrossRef]

97. García-Hernández, J.E.; Hoeksema, B.W. Sponges as secondary hosts for Christmas tree worms at Curaçao. *Coral Reefs* **2017**, *36*, 1243. [CrossRef]

98. Nugues, M.M.; Bak, R.P.M. Differential competitive abilities between Caribbean coral species and a brown alga: A year of experiments and a long-term perspective. *Mar. Ecol. Prog. Ser.* **2006**, *315*, 75–86. [CrossRef]

99. Nugues, M.M.; Bak, R.P.M. Long-term dynamics of the brown macroalga *Lobophora variegata* on deep reefs in Curaçao. *Coral Reefs* **2008**, *27*, 389–393. [CrossRef]

100. Kicklighter, C.E.; Hay, M.E. To avoid or deter: Interactions among defensive and escape strategies in sabellid worms. *Oecologia* **2007**, *151*, 161–173. [CrossRef]

101. Stabili, L.; Schirosi, R.; Di Benedetto, A.; Merendino, A.; Villanova, L.; Giangrande, A. First insights into the biochemistry of *Sabella spallanzanii* (Annelida: Polychaeta) mucus: A potentially unexplored resource for applicative purposes. *J. Mar. Biol. Assoc. U. K.* **2011**, *91*, 199–208. [CrossRef]

102. Giangrande, A.; Licciano, M.; Schirosi, R.; Musco, L.; Stabili, L. Chemical and structural defensive external strategies in six sabellid worms (Annelida). *Mar. Ecol.* **2014**, *35*, 36–45. [CrossRef]

103. Stabili, L.; Schirosi, R.; Licciano, M.; Giangrande, A. Role of *Myxicola infundibulum* (Polychaeta, Annelida) mucus: From bacterial control to nutritional home site. *J. Exp. Mar. Biol. Ecol.* **2014**, *461*, 344–349. [CrossRef]

104. Coutinho, M.C.L.; Teixeira, V.L.; Santos, C.S.G. A review of "Polychaeta" chemicals and their possible ecological role. *J. Chem. Ecol.* **2018**, *44*, 72–94. [CrossRef]

105. Barton, J.A.; Bourne, D.G.; Humphrey, C.M.; Hutson, K.S. Parasites and coral-associated invertebrates that impact coral health. *Rev. Aquac.* **2020**, *12*, 2284–2303. [CrossRef]

106. Adhavan, D.; Prakash, S.; Kumar, A. Tube dwelling gastropod an indicator of coral reef status at the tropical reef of Palk Bay region, southeast coast of India. *Ind. J. Geo Mar. Sci.* **2021**, *50*, 585–587.

107. Liu, J.C.W.; Hoeg, J.T.; Chan, B.K.K. How do coral barnacles start their life in their hosts? *Biol. Lett.* **2016**, *12*, 20160124. [CrossRef]

108. Yu, M.-C.; Dreyer, N.; Kolbasov, G.A.; Høeg, J.T.; Chan, B.K.K. Sponge symbiosis is facilitated by adaptive evolution of larval sensory and attachment structures in barnacles. *Proc. R. Soc. B* **2020**, *287*, 20200300. [CrossRef]

 *diversity*

MDPI

Article

The Association of *Waminoa* with Reef Corals in Singapore and Its Impact on Putative Immune- and Stress-Response Genes

Giorgia Maggioni [1,2,3], Danwei Huang [3,4], Davide Maggioni [1,2], Sudhanshi S. Jain [3], Randolph Z. B. Quek [3,4], Rosa Celia Poquita-Du [3], Simone Montano [1,2], Enrico Montalbetti [1,2] and Davide Seveso [1,2,*]

[1] Department of Environmental and Earth Sciences (DISAT), University of Milano—Bicocca, Piazza della Scienza 1, 20126 Milano, Italy; g.maggioni8@campus.unimib.it (G.M.); davide.maggioni@unimib.it (D.M.); simone.montano@unimib.it (S.M.); enrico.montalbetti@unimib.it (E.M.)
[2] MaRHE Center (Marine Research and High Education Centre), Magoodhoo Island, Faafu Atoll 12030, Maldives
[3] Department of Biological Sciences, National University of Singapore, Singapore 117558, Singapore; huangdanwei@nus.edu.sg (D.H.); jain.sudhanshi@gmail.com (S.S.J.); randolphquek@nus.edu.sg (R.Z.B.Q.); poquitadurc@nus.edu.sg (R.C.P.-D.)
[4] Tropical Marine Science Institute, National University of Singapore, Singapore 119223, Singapore
* Correspondence: davide.seveso@unimib.it; Tel.: +39-0264482953

Citation: Maggioni, G.; Huang, D.; Maggioni, D.; Jain, S.S.; Quek, R.Z.B.; Poquita-Du, R.C.; Montano, S.; Montalbetti, E.; Seveso, D. The Association of *Waminoa* with Reef Corals in Singapore and Its Impact on Putative Immune- and Stress-Response Genes. *Diversity* **2022**, *14*, 300. https://doi.org/10.3390/d14040300

Academic Editor: Michael Wink

Received: 2 March 2022
Accepted: 12 April 2022
Published: 15 April 2022

Publisher's Note: MDPI stays neutral with regard to jurisdictional claims in published maps and institutional affiliations.

Abstract: *Waminoa* spp. are acoel flatworms mainly found as ectosymbionts on scleractinian corals. Although *Waminoa* could potentially represent a threat to their hosts, not enough information is available yet regarding their ecology and effect on the coral. Here, the *Waminoa* sp.–coral association was analyzed in Singapore reefs to determine the prevalence, host range, and preference, as well as the flatworm abundance on the coral surface. Moreover, the impact of *Waminoa* sp. on the expression of putative immune- and stress-response genes (*C-type lectin*, *C3*, *Hsp70* and *Actin*) was examined in the coral *Lobophyllia radians*. The association prevalence was high (10.4%), especially in sites with lower sedimentation and turbidity. *Waminoa* sp. showed a wide host range, being found on 17 coral genera, many of which are new association records. However, only few coral genera, mostly characterized by massive or laminar morphologies appeared to be preferred hosts. *Waminoa* sp. individuals displayed variable patterns of coral surface coverage and an unequal distribution among different host taxa, possibly related to the different coral growth forms. A down-regulation of the expression of all the analyzed genes was recorded in *L. radians* portions colonized by *Waminoa* individuals compared to those without. This indicated that *Waminoa* sp. could affect components of the immune system and the cellular homeostasis of the coral, also inhibiting its growth. Therefore, *Waminoa* sp. could represent a potential further threat for coral communities already subjected to multiple stressors.

Keywords: *Waminoa* sp.; association prevalence; Singapore; gene expression; complement pathway; cellular homeostasis

1. Introduction

Scleractinian corals are known to host a variety of organisms belonging to different phyla [1–6]. Among coral ectosymbionts, the acoel flatworms of the genus *Waminoa* (Order Acoela, Family Convolutidae) have been studied only recently. Formerly included in the Platyhelminthes and now placed within the Acoelomorpha [7], *Waminoa* is considered an enigmatic group since much about its ecology and diversity remains unknown [8]. *Waminoa* spp. are found mainly on scleractinian corals, but also on octocorals, sea anemones, corallimorpharians, zoantharians, and echinoderms [8–15], and they show a circumtropical distribution, with the exception of the Caribbean [16]. *Waminoa* flatworms are characterized by the presence of intracellular dinoflagellate symbionts, which have also been found in the worm oocytes, suggesting that these symbionts are inherited via a vertical transmission and not obtained from the host coral [8–10].

These flatworms subsist via different ways, by ingesting the host coral mucus [11,17] and/or by "standing" on the polyps of host anthozoans to obtain floating zooplankton [18]. For these reasons, *Waminoa* spp. could represent a threat to corals, although the worms do not appear to consume coral tissue directly [9]. In fact, acoelomorph flatworms may limit the host feeding on zooplankton by competing for the prey and by physically blocking the coral oral disc, possibly resulting in kleptoparasitism [18]. In this regard, *Galaxea fascicularis* polyps infested by worms showed a significant decrease of prey ingestion rates compared to polyps without worms and between 5 to 50% of total prey captured by the polyps was stolen by *Waminoa* individuals [18]. Since the coral mucus layer aids in heterotrophic feeding and represents a protective physiochemical barrier [19], the removal of the mucus by *Waminoa* spp. may reduce the coral's resistance to pathogens and environmental stressors [11,17]. In addition, being able to reach high densities and cover a significant portion of the coral [20], *Waminoa* spp. may also cause light shading affecting the coral's photophysiology, reducing the productivity of the coral holobiont [11,13]. Despite the possible negative effects of *Waminoa* worms on their hosts, specific studies to diagnose the health of the infested corals have never been performed so far.

As sessile organisms, corals rely on the modulation of their cellular and molecular mechanisms as the first defensive line against environmental stresses and invading pathogens, and these mechanisms represent useful biomarkers of corals' health status [21–24]. In this context, corals possess an innate immune system consisting of different self/non-self-recognition receptors, which activate specialized cellular and humoral signaling pathways leading to diverse downstream effector responses [25,26]. Among the complex network of immune processes, the complement pathway is a proteolytic cascade, by which pattern recognition receptors (PPRs), such as lectins, initiate intracellular signaling to enact complement component factors, such as C3-like proteins [27]. Lectins are recognition receptors that bind to glycans and play a role in non-self-recognition, cell-cell adhesion, bacterial cell wall recognition, and phagocytosis [28]. In particular, C-type lectins are a superfamily of Ca^{2+}-dependent carbohydrate-recognition proteins involved in the activation of several innate immune responses [29–31]. Complement C3-like proteins, whose activation relies on lectins, are important in allorecognition and involved in the opsonization of the pathogens, chemotaxis, and activation of leukocytes [25,27]. Although complement-encoding genes, such as those of C-type lectins and C3, have been identified and characterized in corals [32–36], their involvement and modulation in response to biotic stressors remain poorly tested.

In addition to the immune response components, other diagnostic tools in corals able to reflect changes in cellular integrity and functionality caused by stress exposure are cellular proteins, such as Actin and Heat shock proteins. Indeed, Actin, which is a major cytoskeletal protein involved in cell motility, growth, and division, is thought to be a proxy of the growth rate in corals [37,38]. Heat shock proteins (Hsps) are molecular chaperones involved in cytoprotection and maintenance of protein homeostasis, and their expression is usually up-regulated when organisms face conditions that may affect their cellular protein structure [39]. For this reason, Hsps have been frequently adopted as cellular stress biomarkers in corals subjected to different environmental stressors [40–45]. In addition, Hsps may play a role in the coral immune system, since they can be activated in response to epizootic diseases or other biotic stresses [46–49].

The coral reefs of Singapore, an island megacity that has been experiencing intense urban development over the past 60 years, represent highly disturbed and urbanized coastal environments [50,51]. Coral communities here have been affected for decades by multiple chronic anthropogenic pressures, resulting in high levels of sedimentation and turbidity [52–55], and multiple bleaching events caused by climate change [56–58]. However, no work has examined the presence and the impact of *Waminoa* worms on Singapore's coral communities.

In this study, the association between *Waminoa* and scleractinian hosts was studied for the first time in Singapore reefs through an ecological survey and a molecular analysis.

In particular, we determined the prevalence of the association, the host range, the host preference of *Waminoa*, and the flatworm abundance on the coral surface. In addition, we tested a possible effect of *Waminoa* on components of the coral immune system and cellular stress response. For this purpose, the coral *Lobophyllia radians* was selected being a species highly colonized by the *Waminoa* in Singapore, and the gene expression of *C-type lectin*, *C3*, *Hsp70* and *Actin* were examined and used as biomarkers to provide new insights on the nature of the association.

2. Materials and Methods

2.1. Ecological Analysis

2.1.1. Study Area and Sampling Design

Fringing reefs of two islands south of mainland Singapore, Pulau Hantu (1°22′74″ N, 103°74′65″ E) and Kusu Island (1°22′55″ N, 103°85′85″ E), were selected as study sites (Figure 1). Both reefs are characterized by a shore-adjacent reef flat leading seaward to the reef crest and down the reef slope to ~8–10 m maximum depth because of high levels of suspended sediments that cause extreme light attenuation [55,59].

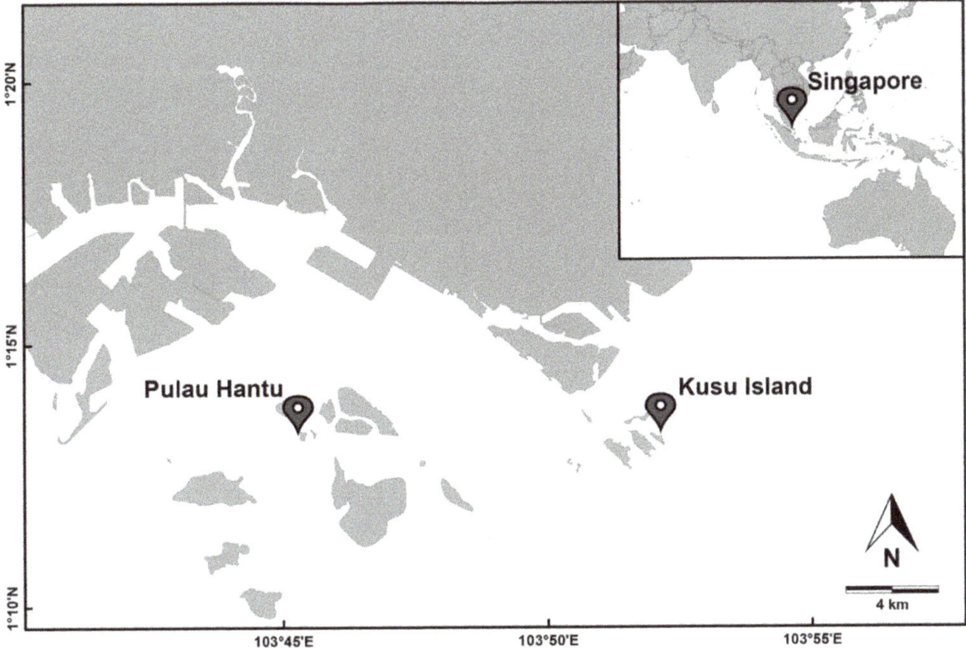

Figure 1. Map of the study area showing the two investigated sites. Pulau Hantu is a sheltered site situated 8 km south of mainland Singapore, while Kusu Island is located 6.4 km south of Singapore's city center and 13.4 km east of Pulau Hantu.

In both sites, extensive surveys were conducted by SCUBA diving between August and October 2019 to detect the occurrence of *Waminoa* individuals on different coral genera (Figure 2). Images of *Waminoa* specimens were taken by a digital camera and analyzed. We distinguished a single morphotype of *Waminoa* in the surveyed area (as described in [8,15], Figure 2) and we treated it as *Waminoa* sp.

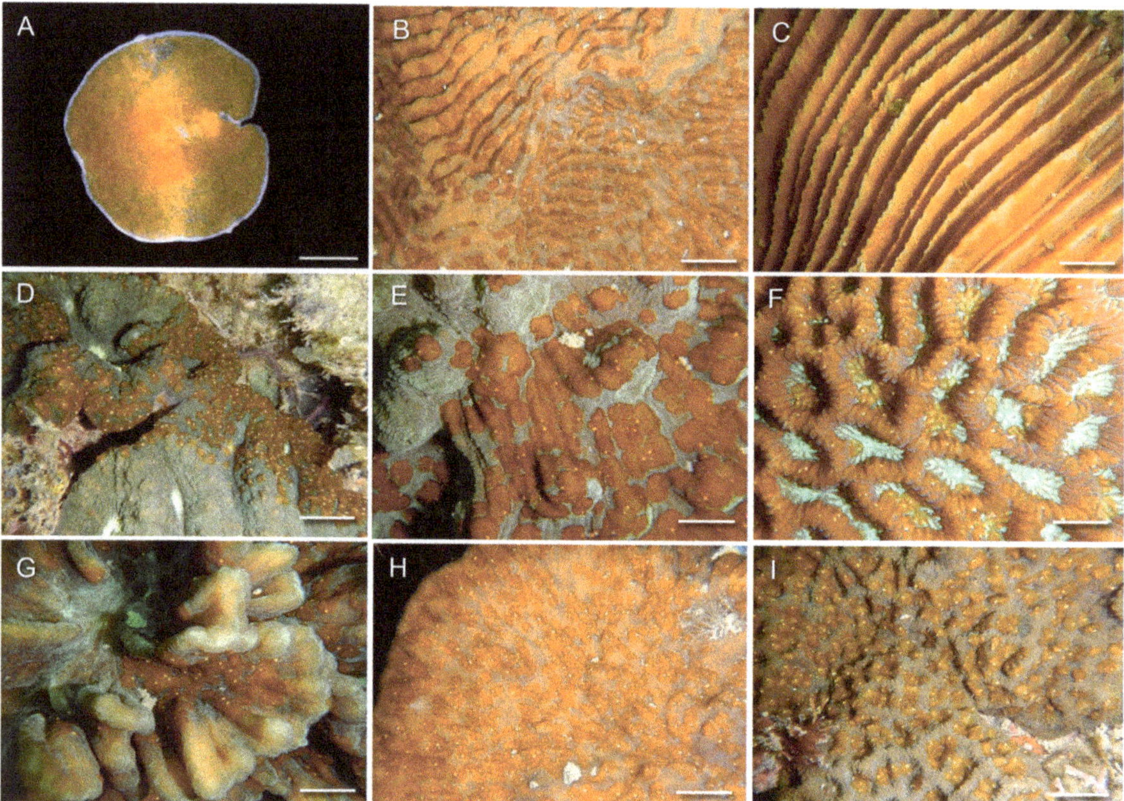

Figure 2. *Waminoa* sp. individuals in association with different coral genera in Singapore reefs. (**A**) Close up of a single *Waminoa* sp. individual characterized by an obcordate general shape, body with brown coloration, white peripheral outer edge, white random dots, and white comparatively larger one white internal spot. Flatworms colonizing corals belonging to the genera *Pachyseris* (**B**), *Fungia* (**C**), *Lobophyllia* (**D**,**E**), *Goniastrea* (**F**), *Pectinia* (**G**), *Merulina* (**H**), and *Favites* (**I**). Scale bars: 1 mm for (**A**) ~1 cm for (**C**,**E**,**F**,**G**) ~2 cm for (**B**,**D**,**H**,**I**).

Six 50 × 2 m belt transects (100 m^2 each), spaced 10 to 20 m apart and placed parallel to the coast at a constant depth between 5 and 8 m (depending on the tides) along the reef slope, were randomly laid at each site. Within belt transects all the coral colonies were identified to genus level, according to Huang et al. [60] and Wong et al. [61], and the number of colonies of each genus found colonized by *Waminoa* sp. was recorded. Moreover, in coral colonies hosting *Waminoa* individuals, their abundance was estimated by determining the percentage of coral surface covered by flatworms and was indicated with four coverage categories, as suggested in [62]: low (coral surface covered 1–10% by worms), moderate (11–25%), severe (26–50%), and extreme (>50%).

In each belt transect, the point intercept transect (PIT) method was also performed (by recording data every 10 cm [63]) to determine the composition and structure of the benthic community, as well as the cover percentage of each coral genus. Data were collected using the following benthic categories: algae, dead coral, coral, coral rubble, rock, sand, and other (sponges, soft corals, tunicates, zoantharians, and unknown). Furthermore, for the macro-category "coral", the genus was also recorded.

2.1.2. Data Analysis

For each transect, the prevalence of the association was calculated as the ratio between the number of corals colonized by *Waminoa* sp. and the total number of colonies. By averaging the corresponding prevalence values measured on the six random belt transects for each site, both an overall and a series of taxon (coral genus)-specific prevalence values were determined. Data normality was verified using the Shapiro–Wilk test. A one-way ANOVA was used to test significant differences in the overall association prevalence between the two sites analyzed. The same analysis, followed by Tukey's honestly significant difference (HSD) post hoc tests for multiple pairwise comparisons of means, was performed to assess significant differences in the prevalence of the *Waminoa* sp.–coral association among the different host genera. Coral genera showing a prevalence < 5% were not included in the analysis.

The host preference of *Waminoa* sp. in terms of coral genus was tested through the Van der Ploeg and Scavia Selectivity coefficient (Ei), following [64]. This coefficient is defined for a group *i* as:

$$ \mathrm{Ei} = \frac{\left[Wi - \left(\frac{1}{n} \right) \right]}{\left[Wi + \left(\frac{1}{n} \right) \right]} $$

where *Wi* represents the value of Chesson's α and *n* represents the number of habitat types [63], here represented by the coral genera found in the study area. Chesson's α value (*Wi*) is defined as:

$$ Wi = \frac{ri}{Pi} / \sum_{i} ri / Pi $$

where *ri* represents the frequency of a habitat category (coral genus) in the environment, and Pi represents the frequency of the same habitat category in which the organism of interest (*Waminoa* sp.) is found [65]. Values of selectivity coefficient range between −1 and 1, with −1 meaning complete avoidance of a host coral genus, and 1 meaning exclusive preference for a specific coral genus [66].

A one-way ANOVA followed by a Tukey's HSD post hoc test was performed to assess differences in the coverage percentages of the flatworms on the surface of the host corals, between the four coverage categories.

2.2. Molecular Analysis

2.2.1. Coral Collection

To assess the effect of *Waminoa* sp. on coral's gene expression, five colonies of *Lobophyllia radians* showing patches of *Waminoa* sp. individuals in a fixed position on their surface were selected, monitored for several days and sampled on Kusu Island. All five colonies were moderately (11–25%) covered by the worms. For each colony, fragments of approximately 2 cm^2 were collected using a hollow-point stainless steel spike [67]. Two coral fragments were sampled from each colony: one fragment located just underneath a patch of *Waminoa* sp. individuals (marked as "W"), and the other from a colony portion without worms, at least 5 cm away from the *Waminoa* individuals (marked as "W/0"). In addition, three healthy colonies of *L. radians* not colonized by the flatworms were randomly sampled as controls to test primer efficiencies.

All coral samples were taken at the same hour (around 9:00 a.m.), at the same shallow depth, and during high tide, to minimize any possible effects of abiotic variables, such as water temperature and light intensity, on the gene expression. During the sampling period, the sea surface temperature was continuously logged and its mean was 30.36 ± 0.39 °C, with very slight oscillation between 29.66 and 30.42 °C. In "W" fragments, *Waminoa* individuals were removed from their hosts by using the pipettes as previously described [8] and the morphological condition of the coral tissues, as well as the presence of any physical damages or lesions, were evaluated. All the coral portions were immediately placed in pre-labeled tubes and, at the end of the underwater sampling, the seawater was decanted and replaced with RNAlater (Thermo Fisher Scientific, Singapore) with a tissue:RNAlater ratio of 1:10. The maximum time between collection and placement in RNAlater was about 30 min. The sample tubes were inverted to mix for 30 s and kept at 4 °C overnight to allow complete penetration of RNAlater into the coral tissues. The tubes were subsequently stored at −80 °C until RNA extraction.

2.2.2. RNA Extraction and REVERSE Transcription (RT)

Total RNA was extracted from all the coral samples without homogenization to reduce RNA fragmentation using TRIzol Reagent (Life Technologies, Sigma-Aldrich, Singapore) following the manufacturer's protocol. RNA quality was checked by examining with gel electrophoresis for presence of clear bands of ribosomal RNAs and RNA concentration was estimated using Qubit (RNA Broad Range Assay Kit, Thermo Fisher Scientific). Complementary DNA (cDNA) was immediately prepared from 1 µg of total RNA for each sample, using a one-tube format of iScript RT Supermix (Bio-Rad, Hercules, CA, USA) for a reverse transcription quantitative polymerase chain reaction (RT-qPCR). Reaction setup was composed of iScript RT Supermix (4 µL), RNA template (varied depending on RNA sample concentration; 14.6–200 ng/µL), and nuclease-free water (variable), with a final volume of 20 µL as previously described [68]. Incubation of the reaction mix was performed in a Labcycler (Sensoquest, Göttingen, Germany) following the manufacturer's protocol: priming for 5 min at 25 °C, RT for 20 min at 45 °C, and RT inactivation for 1 min at 95 °C.

2.2.3. Primer Design and Validation

A local search in the *L. radians* transcriptome previously sequenced (raw sequencing data available at NCBI Sequence Read Archive under accession number PRJNA512601) and assembled by [69] using BLASTn against the GenBank database was performed. Coral transcripts matching genes from the genomes of *Acropora digitifera*, *Orbicella faveolata*, or *Stylophora pistillata* were identified and orthologous genes on the *L. radians* transcriptome selected. The accuracy of the sequence of each gene of interest (GOI) was checked using the NCBI Nucleotide BLAST tool (https://blast.ncbi.nlm.nih.gov/Blast.cgi, accessed on 15 November 2019). A megablast search was performed to ensure that the sequences producing significant alignments with the query sequence were corresponding to the GOIs. A BLASTn search was performed in an open-access coral genomic database at http://reefgenomics.org/blast/ [70,71] with *Goniastrea aspera* genome as subject and the sequence of each GOI as query. A suitable region within exons was selected and the query corresponding to the longer hit with the highest identities value was selected to design primers. Primers for each gene were designed using the online tool by NCBI that incorporated Primer3 and BLAST (https://www.ncbi.nlm.nih.gov/tools/primer-blast). The set of primers given as output by the NCBI tool were analyzed and used in http://reefgenomics.org/blast/to perform a BLASTn search against Symbiodiniaceae nucleotide databases (Supplementary materials Table S1) to verify primer specificity for coral DNA. The selected set of primer was then used for a megablast search using the NCBI Nucleotide BLAST tool to verify the absence of significant alignment with marine species found in the same environment as *L. radians*. The designed primers are shown in Table 1. The specificity of the selected primers for *L. radians* was tested by performing PCR using the GoTaq Green Master Mix in a LifeECO Thermal Cycler (Bioer Technology, Hangzhou, China). PCR reaction steps were (1) denaturation: 95 °C for 45 s, (2) annealing: 50–55 °C for 45 s, and (3) extension: 72 °C for 3 min, repeated for 35 cycles.

Table 1. List of genes of interest (GOIs) with primer designs. The melting temperature (Tm), the % of guanine and cytosine (GC), and the length of PCR product (bp) are also reported for each gene. Additional information of these sequences can be found in Supplementary Materials.

Gene	Sequence	Tm (°C)	GC %	PCR Product
C- type lectin	F: 5′–GTT CTA CTG GGT AGA CGA CA–3′ R: 5′–GAA CAT CAT TCC ATG GTC CC–3′	53.2 53.4	50.00 50.00	155 bp
C3	F: 5′–GTT GAG TTC CCT GAT GCA AT–3′ R: 5′–CAA CAG GTA AAC GCT TTG G–3′	50.9 52.0	40.00 47.37	159 bp
Hsp70	F: 5′–ACA ACT CCC AGC TAT GTC GC–3′ R: 5′–TCC ACT CTC CCT TGG TCT GT–3′	57.3 57.6	55.00 55.00	226 bp
Actin	F: 5′–ATG GTT GGT ATG GGT CAG AAA G–3′ R: 5′–TCT GTT AGC TTT TGG GTT GAG T–3′	54.8 54.3	45.45 40.91	219 bp

To test primer efficiencies, a series of twofold dilutions of *L. radians* cDNA starting from 5 ng/µL were performed for the samples collected from the control colonies [38]. Each dilution was used in triplicate for each primer to assess the primer efficiency through an RT-qPCR using the CFX96 Real-Time PCR System (Bio-Rad). Calculations of efficiencies (E, the amplification factor per PCR cycle) needed to correct for amplification efficiencies per primer were undertaken using an MCMC.qpcr package in R developed by Matz et al. [72]. The function, PrimEff( ), calculates E and plots the regression slopes and E based on dilution series. GOIs with E values outside the 1.85–2.16 range had primers redesigned and re-validated. GOIs that failed amplification were excluded from downstream analyses [68].

2.2.4. Gene Expression Quantification

RT-qPCRs were performed in a CFX96TM Real-Time PCR System (Bio-Rad) using SsoAdvanced inhibitor-tolerant SYBR Green Supermix following the manufacturer's protocol (polymerase activation and DNA denaturation: 3 min at 98 °C, denaturation: 15 s at 95 °C, and annealing/extension: 30 s at 60 °C) repeated for 40 cycles. The reaction mix was prepared as in Table S2. Each sample was tested in duplicates for each of the four genes. To control for variations in expressions of genes caused by differences in RNA concentration of each sample, the amount of cDNA template was standardized to ~10 ng of cDNA for every reaction mix.

2.2.5. Data Analysis

Data obtained from RT-qPCRs expressed as "cycle of quantification values" (i.e., Cq values) were collated and sorted for subsequent analysis. RT-qPCR data were analyzed using generalized linear mixed models based on lognormal-Poisson error distribution, fitted using the MCMC.qpcr package Version 1.2.3 in R Studio Version 1.1.463 as previously reported [72]. Molecule count data with corrections for primer efficiencies were derived with amplification efficiencies (E) per gene and Cq for a single target molecule using the formula: Count = $E(Cq1 - Cq)$. The Cq-to-counts conversion is the key transformation in this method, in which higher variation at the low gene expression values is properly accounted for by the relative quantification model. The transformation makes it possible to fit the resulting data to generalized linear mixed models to account for Poisson-distributed fluctuations when the number of the molecule count is low. Similar Bayesian approaches for analyzing qPCR data have been used in several other reports [68]. Results of the mcmc.qpcr() function were then plotted with HPDsummary() to visualize fold changes in gene expression in response to the presence of *Waminoa* worms. HPDsummary() also calculates all the pairwise differences between treatments and their statistical significance for each gene. Each gene profile was examined to determine the differential expression level between coral samples underneath the surface of the coral colonized by *Waminoa* sp. and samples of the same colonies at least 5 cm apart from the flatworms.

A multivariate analysis was performed using the statistical package PRIMER-E v.7 with the PERMANOVA+ add on [73,74] to investigate together the modulation of all biomarkers in response to *Waminoa* sp. colonization. In particular, data related to the levels of all the biomarkers were square root transformed to calculate a matrix based on the Bray–Curtis similarity. To test for differences in biomarker levels between corals with and without worms, a non-parametric permutational multivariate analysis of variance (PERMANOVA) was performed using 999 permutations with partial sum of squares and unrestricted permutation of raw data. Values were considered statistically significant at $p < 0.05$.

3. Results

3.1. Ecological Analysis

On the surveyed reefs, benthic coverage was dominated by hard corals (36 ± 8.6%), followed by dead corals (23.3 ± 16.1%) and coral rubble (20 ± 6.4%), and the same trend was observed in both sites (Figure S1). A total of 39 scleractinian genera were recorded (Table S3) and among them the genus *Pectinia* showed the highest cover percentage (~11%), followed by *Dipsastraea* and *Merulina* (~9% and 8%, respectively). All the other coral genera displayed a cover percentage close to or less than 5% (Table S3). However, the two sites at Pulau Hantu and Kusu Island showed a distinct abundance and diversity of the various coral genera (Table S3). Indeed, in Pulau Hantu 33 genera were observed, while Kusu Island showed a higher coral diversity with 38 genera recorded (Table S3). In addition, Pulau Hantu reef was dominated by corals belonging to the genera *Pectinia* (~13%), *Merulina* (~10%), *Goniopora*, and *Dipsastraea* (both ~8%), while Kusu Island displayed a greater heterogeneity in terms of coral genera coverage, with *Dipsastraea* (~10%), *Heliopora* and *Pectinia* (all ~8%), *Favites* (~7%), and *Pachyseris*, *Montipora* and *Platygyra* (all ~6%) showing the highest abundance (Table S3).

Overall, 1044 coral colonies were observed, and the overall prevalence of the coral–*Waminoa* sp. association was 10.4 ± 2.3%, with the site on Kusu Island showing a significantly higher prevalence compared to Pulau Hantu (ANOVA, F(1,34) = 3.1, $p = 0.012$; Figure 3). In total, 17 out of the 39 scleractinian genera recorded in the study area were found in association with *Waminoa* sp. and significant differences in the association prevalence were detected among the host coral genera (ANOVA, F(12,143) = 2.39, $p = 0.008$; Figure 4A). In particular, *Lobophyllia* clearly displayed the highest prevalence, followed by *Goniastrea* and *Favites*, and later by *Mycedium*, *Platygyra*, *Oxypora*, and *Pachyseris*, while for the five other host genera the prevalence recorded was lower than 10% but higher than 5% (*Pectinia*, *Echinopora*, *Fungia*, *Ctenactis*, and *Podabacia*, Figure 4A). However, significant differences in the association prevalence were observed only between *Lobophyllia* and *Echinopora*, and *Fungia*, *Ctenactis* and *Podabacia* (Figure 4A). In addition, the five other scleractinian genera, namely *Merulina*, *Porites*, *Dipsastraea*, *Hydnophora*, and *Montipora*, were found associated with *Waminoa* sp. but with prevalence < 5% (4.2, 3.7, 3.5, 2.5, and 1.9%, respectively). The prevalence patterns recorded in both sites were not uniform and did not fully reflect those recorded in the whole study, with *Goniastrea*, *Pachyseris* and *Lobophyllia* showing the higher prevalence in Pulau Hantu, and *Lobophyllia*, *Favites*, *Mycedium*, and *Oxypora* in Kusu Island (Figure S2). The selectivity coefficient Ei allowed the comparison of the relative abundance of coral genera colonized by *Waminoa* sp. with the relative abundance of the same coral genera recorded in the whole study area (Figure 4B). The analysis was performed only for coral genera showing an association prevalence > 5%. It revealed that *Mycedium* was the preferred host for *Waminoa* sp., followed by *Lobophyllia*, *Oxypora*, and, surprisingly, *Ctenactis*, which was among the coral genera that showed the lowest prevalence of the association (Figure 4B). Moreover, *Waminoa* sp. showed a marked avoidance for high/medium-prevalence genera, such as *Platygyra*, *Pachyseris* and *Pectinia* (Figure 4B).

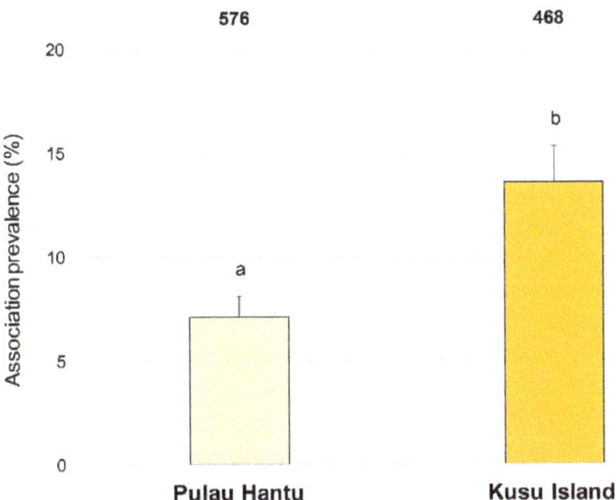

Figure 3. Prevalence (%) of *Waminoa* sp.–corals associations in Pulau Hantu and Kusu Island. Numbers above each bar indicate the total number of coral colonies (both with and without *Waminoa*) analyzed per site. Data are expressed as the mean ± SEM. One-way ANOVA was performed between sites. Letters on the bars denote significant difference among sites. Different letters indicate significant difference ($p < 0.05$), while same letter indicates no significant difference ($p \geq 0.05$).

Most of the coral colonies were moderately (from 11 to 25% of the coral surface) or severely (26–50%) covered by *Waminoa* sp., while a few colonies showed an extreme colonization of worms (>50%) on their surface (Figure 5A). However, no significant differences in the abundance percentages were recorded among the *Waminoa* sp. coverage categories (ANOVA, $F(3, 44) = 1.83$, $p = 0.158$, Figure 5A). Therefore, the distribution of the flatworms on the coral colonies was heterogeneous, although *Waminoa* sp. mostly occupied less than 50% of the coral's surface (Figure 5A). This pattern was also found in the coral genera showing the highest prevalence of the association, such as *Lobophyllia*, *Goniastrea*, and *Favites*, as well as in the preferred genus *Mycedium* (Figure 5B). In particular, in *Lobophyllia* corals the flatworms mostly colonized from 26 to 50% of the colony surface, while in both *Goniastrea* and *Favites* about 40% of the colonies had less than 10% of their area occupied by *Waminoa* sp. individuals. Almost all the colonies of *Mycedium* hosting the worms had less than 10% of their surface covered, while the *Waminoa* sp. infestation on the coral surface was mostly severe in *Pachyseris* and extreme in *Fungia* (Figure 5B).

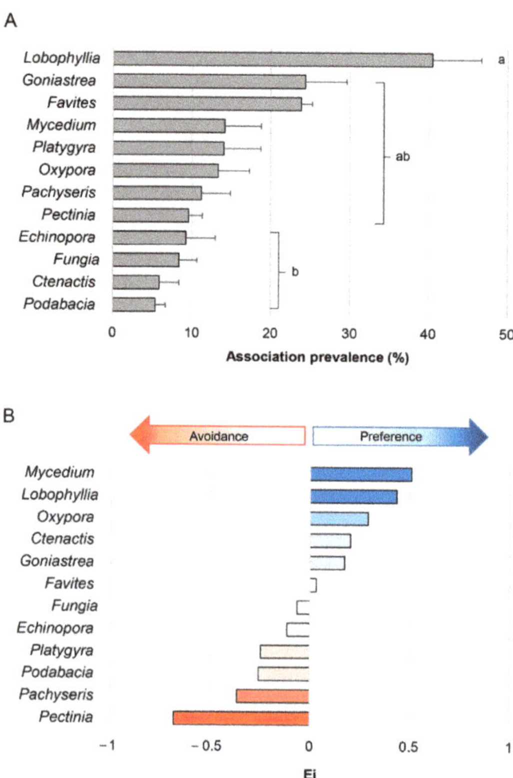

Figure 4. (**A**) Prevalence (%) of *Waminoa* sp.–corals associations by genus in the whole study area. Data are expressed as mean $\pm$ SEM. Letters denote Tukey's significant differences among the different groups ($p < 0.05$); the same letter indicates no significant difference ($p \geq 0.05$). (**B**) Host preferences of *Waminoa* sp. according to the Van der Ploeg and Scavia selectivity coefficient Ei (-1 = complete avoidance; 0 = random choice; +1 = exclusive preference) for each coral genus. In both graphs, only the coral genera with an association prevalence > 5% are reported.

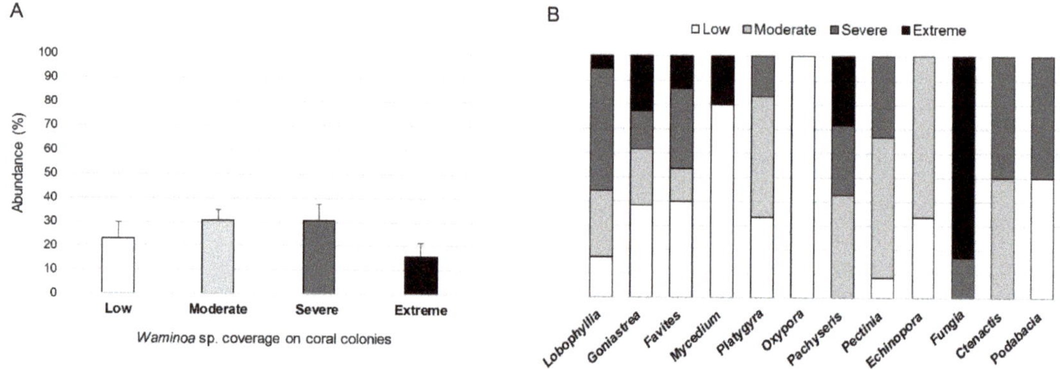

Figure 5. (**A**) Abundance (%; mean $\pm$ SE) of the coral colonies in association with *Waminoa* individuals based on the surface coverage (categories are explained in the Materials and Method section) by the flatworms. (**B**) Relative abundance (%) of coral colonies per genus showing a low, moderate, severe, or extreme distribution of *Waminoa* sp. individuals on their surface.

3.2. Molecular Analysis

All the sampled coral portions occupied by worms showed no visible surface damage or lesions. All the candidate genes showed reliable amplification, since the efficiency of amplification was within the range of acceptable values of 1.49–2.2 (Table S4, [38]).

Significant differences in biomarker levels among portions of coral tissue colonized or not by *Waminoa* sp. were recorded (PERMANOVA: df = 1, F = 3.372, *p* = 0.007). All the genes showed a lower expression level in samples collected underneath *Waminoa* sp. compared to samples of the same coral colony taken at least 5 cm apart from the flatworms (Figure 6A). Therefore, the presence of *Waminoa* sp. on coral caused a down-regulation of the expression of all the investigated genes in the portion of coral tissue directly in contact with the flatworms. The effect of the presence of *Waminoa* sp. on the gene expression was significant for *C3* (pMCMC = 0.03), *Hsp70* (pMCMC = 0.01), and *Actin* (pMCMC = 0.005), but not for *C-type lectin* (pMCMC = 0.13), (Figure 6A). The highest fold change in expression levels was observed for *Actin*, followed by *Hsp70*, while the lowest change was observed for *C-type lectin*, which showed a non-significant down-regulation (Figure 6B).

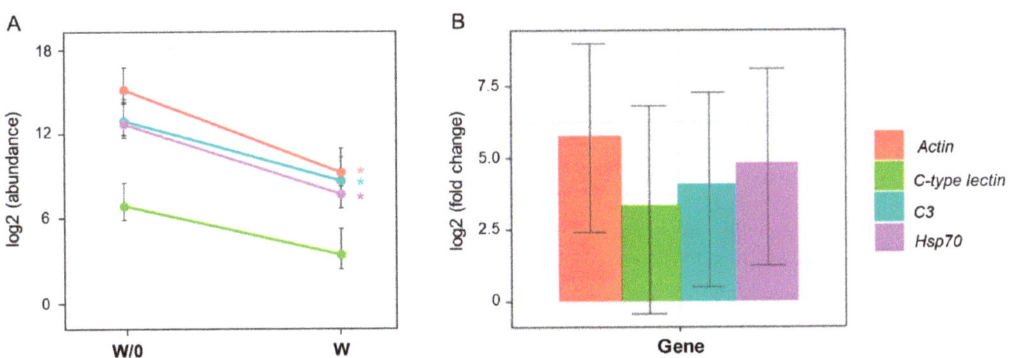

Figure 6. (**A**) Changes in expression levels of the analyzed genes (log2-transformed) between coral fragments not infested (W/0) and infested with *Waminoa* sp. (W). Significant differences in the gene expression abundance are indicated with asterisks. (**B**) Modulation of each gene as fold change. Fold changes were calculated with respect to levels detected in "W" fragments and were log2-transformed. In both graphs, data are expressed as means (*n* = 5).

4. Discussion

4.1. Ecology of the Waminoa-Coral Association in Singapore Reef

Our data contributed to extend the geographic distribution of the *Waminoa*–coral association. In fact, to date it has been recorded only in the reefs of the Red Sea [9,11,12,17], Indonesia [13,20,75], Micronesia [76], Australia [77,78], Japan [8,14], and Taiwan [79]. In addition, our results indicate that the association appeared to be abundant in the study area, with a prevalence greater than 10%. In areas close to Singapore, such as Taiwan, less than 1% of the corals analyzed were found colonized by *Waminoa* sp. [79], while in Wakatobi (Sulawesi, Indonesia) a total of 4.8% of all observed hard corals were associated with the acoel worm in 2006 and 2.6% of hard and soft corals in 2007 [13]. However, a comparison between the prevalence obtained in these studies may not be completely reliable, since the investigated geographic areas were characterized by diverse habitat structure and ecological traits, and the survey methods and approaches used were not entirely the same.

Although the presence of the flatworms was reported in both the investigated sites, a significantly higher prevalence of the association was observed on Kusu Island compared to Pulau Hantu. This difference could be explained by the greater diversity of coral genera, rugosity, and reef complexity on Kusu Island [59], which may have contributed to available niches for *Waminoa* sp., as well as by the different environmental and physical characteristics of the sites. Pulau Hantu is sheltered by adjacent and heavily developed islands in an area of intense industrialization and ship traffic, while Kusu Island experiences comparatively lower anthropogenic impacts and higher exposure to wave action [80,81]. This generates a significantly higher average turbidity, sedimentation, and light attenuation rate in Palau Hantu, resulting in an overall lower light intensity and shallower euphotic depth than Kusu Island [81]. These conditions potentially affect the presence of photosynthetic dinoflagellate-hosting organisms such as *Waminoa* flatworms, which may also have more difficulty in colonizing sediment-covered surfaces and may themselves be vulnerable to environmental disturbances.

The *Waminoa* sp. host range was updated with additional scleractinian genera, many of which are new records. Indeed, in Singapore, *Waminoa* sp. was in association with 17 coral genera belonging to six families, namely Lobophylliidae, Merulinidae, Agariciidae, Poritidae, Fungiidae, and Acroporidae. Among them, the family Merulinidae was largely the most represented, as also recorded in Taiwan, despite only six coral genera in total being found infested by *Waminoa* sp. [79]. In Sulawesi (Indonesia), the association with *Waminoa* was confirmed for 21 coral taxa (Wakatobi [13]), but in Bangka Island it was recorded for only 4 coral genera (*Gardineroseris, Platygyra, Porites, Turbinaria* [75]). In the Red Sea, 13 coral genera were found infected [11]. In Japan, *Waminoa* individuals were found on 4 coral genera only, namely *Cycloseris, Echinomorpha, Echinophyllia*, and *Pachyseris* [14], and 13 scleractinian hosts all belonging to Lobophylliidae [8]. Therefore, *Waminoa* sp. in Singapore coral reefs showed a wide host range. However, only few coral genera such as *Lobophyllia, Mycedium, Oxypora*, and to a lesser extent *Goniastrea* and *Ctenactis*, appeared to be preferred hosts. These coral genera are characterized mainly by massive/submassive or laminar/encrusting colony morphologies, while corals without *Waminoa* sp. are typically branching or columnar. On the contrary, in previous studies *Waminoa* sp. was predominantly observed on branching *Acropora* and *Stylophora* corals, as well as in the columnar *Tubastrea* [11,13]. The question of why *Waminoa* sp. colonizes and/or prefers only specific coral taxa remains largely unanswered. We hypothesize that the coral skeleton morphology could represent a factor driving the choice of the flatworms, given that some structures could favor protection from predators, allowing the worm to hide. Since coral mucus represents a possible food source for *Waminoa* spp. [17], the different mucus production among different coral taxa could also represent an additional host selection factor. This is even more relevant in Singapore's turbid reefs, as some corals can increase or decrease mucus production as a defense mechanism in response to persistent sediment stress [82–85]. In this regard, it would be interesting to analyze the mucus production and composition of the different coral taxa to explore possible correlations with the *Waminoa* presence. Finally, since *Waminoa* also feed on zooplankton caught by corals [18], the ability of a coral species to capture zooplankton, which is determined by its morphology, coral polyp size, and the type of tentacles and nematocysts (reviewed in [86]), may play an important role in the *Waminoa* host selection. However, in addition to these hypotheses, we cannot exclude that the *Waminoa* individuals analyzed here, albeit being of a single morphotype, did not belong to a single species but represented a complex of cryptic species, each of which specialized in a different host.

Corals of Singapore showed variable patterns of flatworm density, ranging from colonies that were densely and extremely infested to others that were only moderately and sparsely populated, as previously observed [11]. However, as also occurred in Okinawa [14], *Waminoa* sp. individuals were not equally distributed among different host taxa. In particular, in Singapore we detected that different *Waminoa* infestation rates could be related to the coral growth form. Indeed, in corals with massive growth forms (such as *Lobophyllia, Goniastrea, Favites*, and *Platygyra*), *Waminoa* sp. showed a heterogeneous pattern of distribution (but in general < 50% of the coral surface was occupied). Corals with a foliose and/or encrusting growth form (*Mycedium, Podabacia, Oxypora*, and *Echinopora*) were sparsely or moderately covered by flatworms, while Fungidae corals (*Ctenactis* and *Fungia*) appeared extremely colonized by *Waminoa* sp., as also previously observed [20].

4.2. Effect of Waminoa sp. on Coral Putative Immune- and Stress-Response Genes

Waminoa spp. can cause physiological damage to corals by inhibiting photosynthesis, reducing the coral tolerance to environmental stress, and impairing coral respiration and feeding [11,17,20]. Our analysis on coral gene expression produced a detailed description of the early response to stress at cell/tissue level, since changes at the molecular level occur before morphological and physiological impairment appear evident [87,88].

Our results show that *Waminoa* sp. affected the analyzed host molecular pathways associated with the coral's stress tolerance and immunity response, causing a uniform down-regulation of the expression of all the investigated genes. Moreover, this modulation was only observed in the physically undamaged coral tissue portions colonized by *Waminoa* individuals and not in those free from flatworms. This might suggest that, as previously observed in corals infected by bacteria or protozoans [24,46,67], the stress response was confined in a restricted area just below the flatworm, even though polyps are linked together by common tissue in the coral colony.

The complement pathway of the immune system is triggered by lectins binding a pathogen-associated molecule and results in the activation of the complement component factor C3 and C4 [89]. Indeed, in corals, *C-type lectin* and C3 protein were usually up-regulated and activated in response to epizootic diseases [47,48,90,91]. However, their down-regulation may reflect suppression of host immunity, as previously observed for the association between corals and the microalga *Chromera* [92]. Mohamed et al. [92] also suggested that the down-regulation of some PRRs could reflect the host attempting to limit interactions with non-beneficial organisms, since both complement C3 and the *C-type lectin* have been implicated in symbiont recognition and in host-symbiont communication [34]. In addition, the down-regulation of *C-type lectin* and C3 has been observed in corals subjected to temperature/light stress, suggesting that these stresses might compromise the coral's immune defenses and therefore increase the coral's susceptibility to diseases [38,93–95]. Likewise, the decreased expression of the *C-type lectin* and C3 here suggests that the presence of *Waminoa* sp. individuals on coral tissue might interfere with the ability of the whole host coral to respond to the attack of various pathogens and at the same time could make it more vulnerable to environmental stressors. However, while the C3 appeared to be significantly down-regulated by the flatworm presence, the decrease in expression of the *lectin* was not significant. Considering that *C-type lectins* have been shown to respond immediately following an immune challenge [34,90] but may not show any significant response at later times [47], we hypothesize that the observed modulation could be influenced by the sampling times.

The cytoplasmic chaperonin Hsp70 is involved in assembly of newly synthesized proteins and in the refolding of misfolded or aggregated proteins, contributing to the protein transfer to different cellular compartments or to the proteolytic machinery and acting as cellular defensive mechanism [96]. Up-regulation of the *Hsp70* has been proposed as an activator of other components of the coral effector immune systems, such as the prophenoloxidase cascade, in corals infected by pathogens [47]. On the contrary, the down-regulation of Hsp70 in corals reflected the impairment of the cellular defense mechanisms that is due to severe and intolerable stress [97–99], and may indicate a reduced activity of the immune system because of diseases [24]. In addition, since Hsps are ATP-dependent chaperones, the decrease of *Hsp70* expression may be related to the high-energy expenditure necessary to reduce the deleterious effects of *Waminoa* sp. and restore cellular damage. However, it is important to underline that, since the roles of *Hsp70* in organismal function are broad, changes in expression of this gene could also reflect changes in other physiological processes.

Actin was the most responsive gene, showing the greatest down-regulation. In addition to being fundamental for cell motility, contractibility, mitosis, and intracellular transport, Actin is also an important part of the nuclear complex, being required for the transcription of RNA polymerases and in the export of RNAs and proteins from the nucleus [100]. Down-regulation of Actin has previously been observed in corals subjected to thermal stress and acidification [38,68,101,102]. Since Actin is a major cytoskeletal component involved in growth, down-regulation of this gene could be indicative of growth inhibition caused by the presence of *Waminoa* sp. Moreover, the reduced expression of *Actin* may reflect a change in the regulation of gene transcription of proteins involved in cytoskeletal interactions and may imply changes in intracellular transport and cell shape/integrity, as previously suggested [102]. The overall down-regulation of all the analyzed genes may reflect a negative effect of the acoelomate ectosymbiont *Waminoa* sp. on the host coral *L. radians*. However, alternative scenarios should also be considered. For example, it could be possible that *Waminoa* did not cause detectable cellular stress to hosts, preferentially colonizing polyps with reduced defense responses, or interfering with polyp feeding and causing the observed gene down-regulation, as reduced resources would lead to reduced investment in defense.

In conclusion, our study demonstrated that *Waminoa* sp. showed a high prevalence and wide host range in Singapore coral reefs and its distribution patterns were specific to certain scleractinian host genera. Moreover, *Waminoa* sp. could impair both the cellular homeostasis and components of the immune system of the host, thus representing a potential further threat for coral communities living in an area already subjected to multiple stresses, such as sedimentation and light limitation. However, further studies analyzing more genes and biomarkers in different hosts are necessary to have a more complete picture of the association.

Supplementary Materials: The following supporting information can be downloaded at: https://www.mdpi.com/article/10.3390/d14040300/s1, Figure S1: Coverage percentage of the different benthic categories in the two sites analyzed and in the whole study area. Data are expressed as the mean ± SEM; Figure S2: Prevalence (%) of *Waminoa* sp.-corals association by coral genus in the two sites. Data are expressed as the mean ± SEM; Table S1: Symbiodiniaceae nucleotide databases; Table S2: RT-qPCR Mastermix used for determining gene expression and efficiency; Table S3: Coverage percentage of each coral genus in the two sites and in the whole study area; Table S4: Gene efficiency (E) for each analyzed gene (SD: standard deviation)

Author Contributions: G.M. and D.S. wrote the manuscript (original draft preparation); D.H., D.M., S.M. and E.M. reviewed and edited the manuscript; D.M., S.M., E.M. and D.S. analyzed the results; D.H., G.M. and D.S. conceived and designed the research and experiments; D.H., S.S.J. and R.Z.B.Q. secured funding for this research and for all reagents and materials; G.M., S.S.J., R.Z.B.Q. and R.C.P.-D. performed and supervised the lab activities. G.M. and S.S.J. conducted the field activities. All authors have read and agreed to the published version of the manuscript.

Funding: This research is supported by the Temasek Foundation under its Singapore Millennium Foundation Research Grant Programme.

Institutional Review Board Statement: All applicable permits and institutional guidelines required to perform the work were followed. Collections were made under permit NP/RP16–156.

Informed Consent Statement: Not applicable.

Data Availability Statement: The datasets generated during and/or analyzed during the current study are available from the corresponding authors on reasonable request.

Acknowledgments: We are grateful to members of the Reef Ecology Lab, National University of Singapore, for their support throughout this project, both in the field and in the laboratory. Special thanks to Nicholas Yap for his support throughout the project. Thanks to Daisuke Taira, Andrea Leong, and Ng Zhi Sheng for their support during diving activities.

Conflicts of Interest: The authors have no relevant financial or non-financial interest to disclose.

References

1. Bos, A.R. Fishes (Gobiidae and Labridae) associated with the mushroom coral *Heliofungia actiniformis* (Scleractinia: Fungiidae) in the Philippines. *Coral Reefs* **2012**, *31*, 133. [CrossRef]
2. Hoeksema, B.W.; Van der Meij, S.E.T.; Fransen, C.H. The mushroom coral as a habitat. *J. Mar. Biol. Assoc.* **2012**, *92*, 647–663. [CrossRef]

3. Bos, A.R.; Hoeksema, B.W. Cryptobenthic fishes and co-inhabiting shrimps associated with the mushroom coral *Heliofungia actiniformis* (Fungiidae) in the Davao Gulf, Philippines. *Environ. Biol. Fish* **2015**, *98*, 1479–1489. [CrossRef]

4. Montano, S.; Seveso, D.; Galli, P.; Puce, S.; Hoeksema, B.W. Mushroom corals as newly recorded hosts of the hydrozoan symbiont *Zanclea* sp. *Mar. Biol. Res.* **2015**, *11*, 773–779. [CrossRef]

5. Montano, S.; Fattorini, S.; Parravicini, V.; Berumen, M.L.; Galli, P.; Maggioni, D.; Arrigoni, R.; Seveso, D.; Strona, G. Corals hosting symbiotic hydrozoans are less susceptible to predation and disease. *Proc. R. Soc. B Biol. Sci.* **2017**, *284*, 20172405. [CrossRef]

6. Maggioni., D.; Arrigoni, R.; Seveso, D.; Galli, P.; Berumen, M.L.; Denis, V.; Hoeksema, B.W.; Huang, D.; Manca, F.; Pica, D.; et al. Evolution and biogeography of the *Zanclea*-Scleractinia symbiosis. *Coral Reefs* **2020**, 1–17. [CrossRef]

7. Baguñà, J.; Riutort, M. Molecular phylogeny of the Platyhelminthes. *Can. J. Zool.* **2014**, *82*, 168–193. [CrossRef]

8. Kunihiro, S.; Reimer, J.D. Phylogenetic analyses of *Symbiodinium* isolated from *Waminoa* and their anthozoan hosts in the Ryukyus Archipelago, southern Japan. *Symbiosis* **2018**, *76*, 253–264. [CrossRef]

9. Ogunlana, M.V.; Hooge, M.D.; Tekle, Y.I.; Benayahu, Y.; Barneah, O.; Tyler, S. *Waminoa brickneri* n. sp. (Acoela: Acoelomorpha) associated with corals in the Red Sea. *Zootaxa* **2005**, *14*, 1–14. [CrossRef]

10. Hikosaka-Katayama, T.; Koike, K.; Yamashita, H.; Hikosaka, A.; Koike, K. Mechanisms of maternal inheritance of dinoflagellate symbionts in the acoelomorph worm *Waminoa litus*. *Zool. Sci.* **2012**, *29*, 559–567. [CrossRef]

11. Barneah, O.; Brickner, I.; Hooge, M.; Weis, V.M.; Benayahu, Y. First evidence of maternal transmission of algal endosymbionts at an oocyte stage in a triploblastic host, with observations on reproduction in *Waminoa brickneri* (Acoelomorpha). *Invert. Biol.* **2007**, *126*, 113–119. [CrossRef]

12. Barneah, O.; Ben-Dov, E.; Benayahu, Y.; Brickner, I.; Kushmaro, A. Molecular diversity and specificity of acoel worms associated with corals in the Gulf of Eilat (Red Sea). *Aquat. Biol.* **2012**, *14*, 277–281. [CrossRef]

13. Haapkylä, J.; Seymour, A.S.; Barneah, O.; Brickner, I.; Hennige, S.; Suggett, D.; Smith, D. Association of *Waminoa* sp. (Acoela) with corals in the Wakatobi Marine Park, South-East Sulawesi, Indonesia. *Mar. Biol.* **2009**, *156*, 1021–1027. [CrossRef]

14. Biondi, P.; Masucci, G.D.; Kunihiro, S.; Reimer, J.D. The distribution of reef-dwelling *Waminoa* flatworms in bays and on capes of Okinawa Island. *Mar. Biodiv.* **2019**, *49*, 405–413. [CrossRef]

15. Kunihiro, S.; Farenzena, Z.; Hoeksema, B.W.; Groenenberg, D.S.; Hermanto, B.; Reimer, J.D. Morphological and phylogenetic diversity of *Waminoa* and similar flatworms (Acoelomorpha) in the western Pacific Ocean. *Zoology* **2019**, *136*, 125692. [CrossRef] [PubMed]

16. Barton, J.A.; Bourne, D.G.; Humphrey, C.; Hutson, K.S. Parasites and coral-associated invertebrates that impact coral health. *Rev. Aquacul.* **2020**, *12*, 2284–2303. [CrossRef]

17. Naumann, M.S.; Mayr, C.; Struck, U.; Wild, C. Coral mucus stable isotope composition and labeling: Experimental evidence for mucus uptake by epizoic acoelomorph worms. *Mar. Biol.* **2010**, *157*, 2521–2531. [CrossRef]

18. Wijgerde, T.; Schots, P.; Van Onselen, E.; Janse, M.; Karruppannan, E.; Verreth, J.A.; Osinga, R. Epizoic acoelomorph flatworms impair zooplankton feeding by the scleractinian coral *Galaxea fascicularis*. *Biol. Open* **2012**, *2*, 10–17. [CrossRef]

19. Brown, B.E.; Bythell, J.C. Perspectives on mucus secretion in reef corals. *Mar. Ecol. Prog. Ser.* **2005**, *296*, 291–309. [CrossRef]

20. Hoeksema, B.W.; Farenzena, Z.T. Tissue loss in corals infested by acoelomorph flatworms (*Waminoa* sp.). *Coral Reefs* **2012**, *31*, 869. [CrossRef]

21. Downs, C.A. Cellular diagnostics and its application to aquatic and marine toxicology. In *Techniques in Aquatic Toxicology*; Ostrander, G.K., Ed.; CRC Press: Boca Raton, FL, USA, 2005; Volume 2, pp. 181–208.

22. Mydlarz, L.D.; McGinty, E.S.; Harvell, C.D. What are the physiological and immunological responses of coral to climate warming and disease? *J. Exp. Biol.* **2010**, *213*, 934–945. [CrossRef] [PubMed]

23. Rosic, N.; Kaniewska, P.; Chan, C.K.K.; Ling, E.Y.S.; Edwards, D.; Dove, S.; Hoegh-Guldberg, O. Early transcriptional changes in the reef-building coral *Acropora aspera* in response to thermal and nutrient stress. *BMC Genom.* **2014**, *15*, 1052. [CrossRef] [PubMed]

24. Seveso, D.; Montano, S.; Reggente, M.A.L.; Maggioni, D.; Orlandi, I.; Galli, P.; Vai, M. The cellular stress response of the scleractinian coral *Goniopora columna* during the progression of the black band disease. *Cell Stress Chap.* **2017**, *22*, 225–236. [CrossRef] [PubMed]

25. Palmer, C.V.; Traylor-Knowles, N. Towards an integrated network of coral immune mechanisms. *Proc. R. Soc. B Biol. Sci.* **2012**, *279*, 4106–4114. [CrossRef]

26. Traylor-Knowles, N.; Connelly, M.T. What is currently known about the effects of climate change on the coral immune response. *Cur. Clim. Change Rep.* **2017**, *3*, 252–260. [CrossRef]

27. Carroll, M.C. The role of complement and complement receptors in induction and regulation of immunity. *Annu. Rev. Immun.* **1998**, *16*, 545–568. [CrossRef]

28. Fujita, T. Evolution of the lectin–complement pathway and its role in innate immunity. *Nat. Rev. Immun.* **2002**, *2*, 346–353. [CrossRef]

29. Christophides, G.K.; Zdobnov, E.; Barillas-Mury, C.; Birney, E.; Blandin, S.; Blass, C.; Brey, P.T.; Collins, F.H.; Danielli, A.; Dimopoulos, G.; et al. Immunity-related genes and gene families in *Anopheles gambiae*. *Science* **2002**, *298*, 159–165. [CrossRef]

30. Ling, E.; Yu, X.Q. Cellular encapsulation and melanization are enhanced by immulectins, pattern recognition receptors from the tobacco hornworm *Manduca sexta*. *Dev. Compar. Immun.* **2006**, *30*, 289–299. [CrossRef]

31. Endo, Y.; Nakazawa, N.; Iwaki, D.; Takahashi, M.; Matsushita, M.; Fujita, T. Interactions of ficolin and mannose-binding lectin with fibrinogen/fibrin augment the lectin complement pathway. *J. Innate Immun.* **2010**, *2*, 33–42. [CrossRef]

32. Dishaw, L.J.; Smith, S.L.; Bigger, C.H. Characterization of a C3-like cDNA in a coral: Phylogenetic implications. *Immunogenetics* **2005**, *57*, 535–548. [CrossRef] [PubMed]

33. Miller, D.J.; Hemmrich, G.; Ball, E.E.; Hayward, D.C.; Khalturin, K.; Funayama, N.; Agata, K.; Bosch, T.C. The innate immune repertoire in Cnidaria-ancestral complexity and stochastic gene loss. *Genome Biol.* **2007**, *8*, R59. [CrossRef] [PubMed]

34. Kvennefors, E.C.E.; Leggat, W.; Kerr, C.C.; Ainsworth, T.D.; Hoegh-Guldberg, O.; Barnes, A.C. Analysis of evolutionarily conserved innate immune components in coral links immunity and symbiosis. *Dev. Comp. Immun.* **2010**, *34*, 1219–1229. [CrossRef] [PubMed]

35. Shinzato, C.; Shoguchi, E.; Kawashima, T.; Hamada, M.; Hisata, K.; Tanaka, M.; Fujie, M.; Fujiwara, M.; Koyanagi, R.; Ikuta, T.; et al. Using the *Acropora digitifera* genome to understand coral responses to environmental change. *Nature* **2011**, *476*, 320–323. [CrossRef]

36. Ocampo, I.D.; Zárate-Potes, A.; Pizarro, V.; Rojas, C.A.; Vera, N.E.; Cadavid, L.F. The immunotranscriptome of the Caribbean reef-building coral *Pseudodiploria strigosa*. *Immunogenetics* **2015**, *67*, 515–530. [CrossRef]

37. Perrin, B.J.; Ervasti, J.M. The actin gene family: Function follows isoform. *Cytoskeleton* **2010**, *67*, 630–634. [CrossRef]

38. Kenkel, C.D.; Aglyamova, G.; Alamaru, A.; Bhagooli, R.; Capper, R.; Cunning, R.; deVillers, A.; Haslun, J.A.; Hédouin, L.; Keshavmurthy, S.; et al. Development of gene expression markers of acute heat-light stress in reef-building corals of the genus Porites. *PLoS ONE* **2011**, *6*, e26914. [CrossRef]

39. Balchin, D.; Hayer-Hartl, M.; Hartl, F.U. In vivo aspects of protein folding and quality control. *Science* **2016**, *353*, aac4354. [CrossRef]

40. Downs, C.A.; Fauth, J.E.; Halas, J.C.; Dustan, P.; Bemiss, J.; Woodley, C.M. Oxidative stress and seasonal coral bleaching. *Free Rad. Biol. Med.* **2002**, *33*, 533–543. [CrossRef]

41. Downs, C.A.; Woodley, C.M.; Fauth, J.E.; Knutson, S.; Burtscher, M.M.; May, L.A.; Ostrander, G.K. A survey of environmental pollutants and cellular-stress markers of *Porites astreoides* at six sites in St. John, US Virgin Islands. *Ecotoxicology* **2011**, *20*, 1914–1931.

42. Louis, Y.D.; Bhagooli, R.; Seveso, D.; Maggioni, D.; Galli, P.; Vai, M.; Dyall, S.D. Local acclimatisation-driven differential gene and protein expression patterns of Hsp70 in *Acropora muricata*: Implications for coral tolerance to bleaching. *Mol. Ecol.* **2020**, *29*, 4382–4394. [CrossRef] [PubMed]

43. Seveso, D.; Montano, S.; Strona, G.; Orlandi, I.; Galli, P.; Vai, M. Exploring the effect of salinity changes on the levels of Hsp60 in the tropical coral *Seriatopora caliendrum*. *Mar. Environ. Res.* **2013**, *90*, 96–103. [CrossRef] [PubMed]

44. Seveso, D.; Montano, S.; Strona, G.; Orlandi, I.; Galli, P.; Vai, M. Hsp60 expression profiles in the reef-building coral *Seriatopora caliendrum* subjected to heat and cold shock regimes. *Mar. Environ. Res.* **2016**, *119*, 1–11. [CrossRef] [PubMed]

45. Montalbetti, E.; Biscéré, T.; Ferrier-Pagès, C.; Houlbrèque, F.; Orlandi, I.; Forcella, M.; Galli, P.; Vai, M.; Seveso, D. Manganese benefits heat-stressed corals at the cellular level. *Front. Mar. Sci.* **2021**, *8*, 803. [CrossRef]

46. Seveso, D.; Montano, S.; Strona, G.; Orlandi, I.; Vai, M.; Galli, P. Up-regulation of Hsp60 in response to skeleton eroding band disease but not by algal overgrowth in the scleractinian coral *Acropora muricata*. *Mar. Environ. Res.* **2012**, *78*, 34–39. [CrossRef] [PubMed]

47. Brown, T.; Bourne, D.; Rodriguez-Lanetty, M. Transcriptional activation of c3 and hsp70 as part of the immune response of *Acropora millepora* to bacterial challenges. *PLoS ONE* **2013**, *8*, e67246. [CrossRef]

48. Libro, S.; Kaluziak, S.T.; Vollmer, S.V. RNA-seq profiles of immune related genes in the staghorn coral *Acropora cervicornis* infected with white band disease. *PLoS ONE* **2013**, *8*, e81821. [CrossRef]

49. Fuess, L.E.; Weil, E.; Mydlarz, L.D. Associations between transcriptional changes and protein phenotypes provide insights into immune regulation in corals. *Dev. Compar. Immun.* **2016**, *62*, 17–28. [CrossRef]

50. Guest, J.R.; Tun, K.; Low, J.; Vergés, A.; Marzinelli, E.M.; Campbell, A.H.; Bauman, A.G.; Feary, D.A.; Chou, L.M.; Steinberg, P.D. 27 years of benthic and coral community dynamics on turbid, highly urbanised reefs off Singapore. *Sci. Rep.* **2016**, *6*, 36260. [CrossRef]

51. Chou, L.M.; Huang, D.; Tan, K.S.; Toh, T.C.; Goh, B.P.L.; Tun, K. World Seas: An Environmental Evaluation. In *The Indian Ocean to the Pacific. World Seas: An Environmental Evaluation*; Sheppard, C.R.C., Ed.; Academic Press: London, UK, 2019; Volume 2, pp. 539–558.

52. Todd, P.A.; Ong, X.; Chou, L.M. Impacts of pollution on marine life in Southeast Asia. *Biodiv. Conserv.* **2010**, *19*, 1063–1082. [CrossRef]

53. Browne, N.K.; Tay, J.K.; Low, J.; Larson, O.; Todd, P.A. Fluctuations in coral health of four common inshore reef corals in response to seasonal and anthropogenic changes in water quality. *Mar. Environ. Res.* **2015**, *105*, 39–52. [CrossRef] [PubMed]

54. Heery, E.C.; Hoeksema, B.W.; Browne, N.K.; Reimer, J.D.; Ang, P.O.; Huang, D.; Friess, D.A.; Chou, L.M.; Loke, L.; Saksena-Taylor, P.; et al. Urban coral reefs: Degradation and resilience of hard coral assemblages in coastal cities of East and Southeast Asia. *Mar. Poll. Bull.* **2018**, *135*, 654–681. [CrossRef] [PubMed]

55. Chow, G.S.E.; Chan, Y.K.S.; Jain, S.S.; Huang, D. Light limitation selects for depth generalists in urbanised reef coral communities. *Mar. Environ. Res.* **2019**, *147*, 101–112. [CrossRef] [PubMed]

56. Guest, J.R.; Baird, A.H.; Maynard, J.A.; Muttaqin, E.; Edwards, A.J.; Campbell, S.J.; Chou, L.M. Contrasting patterns of coral bleaching susceptibility in 2010 suggest an adaptive response to thermal stress. *PLoS ONE* **2012**, *7*, e33353. [CrossRef] [PubMed]

57. Chou, L.M.; Toh, T.C.; Toh, K.B.; Ng, C.S.L.; Cabaitan, P.; Tun, K.; Goh, E.; Afiq-Rosli, L.; Taira, D.; Du, R.C.; et al. Differential response of coral assemblages to thermal stress underscores the complexity in predicting bleaching susceptibility. *PLoS ONE* **2016**, *11*, e0159755. [CrossRef] [PubMed]

58. Ng, C.S.L.; Huang, D.; Toh, K.B.; Sam, S.Q.; Kikuzawa, Y.P.; Toh, T.C.; Taira, D.; Chan, Y.K.S.; Hung, L.Z.T.; Sim, W.T.; et al. Responses of urban reef corals during the 2016 mass bleaching event. *Mar. Poll. Bull.* **2020**, *154*, 111111. [CrossRef] [PubMed]

59. Januchowski-Hartley, F.A.; Bauman, A.G.; Morgan, K.M.; Seah, J.C.; Huang, D.; Todd, P.A. Accreting coral reefs in a highly urbanized environment. *Coral Reefs* **2020**, *39*, 717–731. [CrossRef]

60. Huang, D.; Benzoni, F.; Arrigoni, R.; Baird, A.H.; Berumen, M.L.; Bouwmeester, J.; Chou, L.M.; Fukami, H.; Licuanan, W.Y.; Lovell, E.R.; et al. Towards a phylogenetic classification of reef corals: The Indo-Pacific genera *Merulina*, *Goniastrea* and *Scapophyllia* (Scleractinia, Merulinidae). *Zool. Scrip.* **2014**, *43*, 531–548. [CrossRef]

61. Wong, J.S.Y.; Chan, Y.K.S.; Ng, C.S.L.; Tun, K.P.P.; Darling, E.S.; Huang, D. Comparing patterns of taxonomic, functional and phylogenetic diversity in reef coral communities. *Coral Reefs* **2018**, *37*, 737–750. [CrossRef]

62. Raymundo, L.J.; Bruckner, A.W.; Work, T.M.; Willis, B. *Coral Disease Handbook Guidelines For Assessment, Monitoring and Management*; Couch, C.A., Harvell, C.D., Raymundo, L., Eds.; Coral Reef Targeted Research and Capacity Building for Management Program: St. Lucia, QLD, Australia; Coral Gables, FL, USA, 2008.

63. Hill, J.; Wilkinson, C. *Methods for Ecological Monitoring of Coral Reefs*; Australian Institute of Marine Science: Townsville, Australia, 2004.

64. Lechowicz, M.J. The sampling characteristics of electivity indices. *Oecologia* **1982**, *52*, 22–30. [CrossRef]

65. Chesson, J. Measuring preference in selective predation. *Ecology* **1978**, *59*, 211–215. [CrossRef]

66. Van der Ploeg, H.A.; Scavia, D. Calculation and use of selectivity coefficients of feeding: Zooplankton grazing. *Ecol. Model.* **1979**, *7*, 135–149. [CrossRef]

67. Seveso, D.; Montano, S.; Reggente, M.A.L.; Orlandi, I.; Galli, P.; Vai, M. Modulation of Hsp60 in response to coral brown band disease. *Dis. Aquat. Org.* **2015**, *115*, 15–23. [CrossRef] [PubMed]

68. Poquita-Du, R.C.; Goh, Y.L.; Huang, D.; Chou, L.M.; Todd, P.A. Gene expression and photophysiological changes in *Pocillopora acuta* coral holobiont following heat stress and recovery. *Microorganisms* **2020**, *8*, 1227. [CrossRef] [PubMed]

69. Quek, Z.B.R.; Huang, D. Effects of missing data and data type on phylotranscriptomic analysis of stony corals (Cnidaria: Anthozoa: Scleractinia). *Mol. Phylogen. Evol.* **2019**, *134*, 12–23. [CrossRef] [PubMed]

70. Voolstra, C.R.; Miller, D.J.; Ragan, M.A.; Hoffmann, A.; Hoegh-Guldberg, O.; Bourne, D.; Ball, E.E.; Ying, H.; Foret, S.; Takahashi, S.; et al. The ReFuGe 2020 Consortium—Using "omics" approaches to explore the adaptability and resilience of coral holobionts to environmental change. *Front. Mar. Sci.* **2015**, *2*, 68.

71. Liew, Y.J.; Aranda, M.; Voolstra, C.R. Reefgenomics.Org–a repository for marine genomics data. *Database* **2016**, *2016*, baw152. [CrossRef]

72. Matz, M.V.; Wright, R.M.; Scott, J.G. No control genes required: Bayesian analysis of qRT-PCR data. *PLoS ONE* **2013**, *8*, e71448. [CrossRef]

73. Clarke, K.R.; Gorley, R.N. *Getting Started with PRIMER v7*; PRIMER-E Ltd.: Plymouth, UK, 2015.

74. Anderson, M.; Gorley, R.; Clarke, K. *PERMANOVA+ for PRIMER: Guide to Software and Statistical Methods*; PRIMER-E Ltd: Plymouth, UK, 2008.

75. Ponti, M.; Fratangeli, F.; Dondi, N.; Reinach, M.S.; Serra, C.; Sweet, M.J. Baseline reef health surveys at Bangka Island (North Sulawesi, Indonesia) reveal new threats. *Peer J.* **2016**, *4*, e2614. [CrossRef]

76. Trench, R.K.; Winsor, H. *Symbiosis with Dinoflagellates in Two Pelagic Flatworms, Amphiscolops sp. and Haplodiscus sp.*; Balaban Publishers: Philadelphia, PA, USA, 1987.

77. Winsor, L. Marine Turbellaria (Acoela) from North Queensland. *Mem. Queensl. Mus.* **1990**, *28*, 785–800.

78. Cooper, C.; Clode, P.L.; Thomson, D.P.; Stat, M. A flatworm from the genus *Waminoa* (Acoela: Convolutidae) associated with bleached corals in Western Australia. *Zool. Sci.* **2015**, *32*, 465–473. [CrossRef] [PubMed]

79. Huang, C.Y.; Hwang, J.S.; Yamashiro, H.; Tang, S.L. Spatial and cross-seasonal patterns of coral diseases in reefs of Taiwan: High prevalence and regional variation. *Dis. Aquat. Organ.* **2021**, *146*, 145–156. [CrossRef] [PubMed]

80. Chen, P.Y.; Chen, C.C.; Chu, L.; McCarl, B. Evaluating the economic damage of climate change on global coral reefs. *Glob. Environ. Change* **2015**, *30*, 12–20. [CrossRef]

81. Morgan, K.M.; Moynihan, M.A.; Sanwlani, N.; Switzer, A.D. Light limitation and depth-variable sedimentation drives vertical reef compression on turbid coral reefs. *Front. Mar. Sci.* **2020**, *7*, 931. [CrossRef]

82. Stafford-Smith, M.G.; Ormond, R.F.G. Sediment-rejection mechanisms of 42 species of Australian scleractinian corals. *Mar. Fresh Res.* **1992**, *43*, 683–705. [CrossRef]

83. Erftemeijer, P.L.; Riegl, B.; Hoeksema, B.W.; Todd, P.A. Environmental impacts of dredging and other sediment disturbances on corals: A review. *Mar. Poll. Bull.* **2012**, *64*, 1737–1765. [CrossRef]

84. Bessell-Browne, P.; Negri, A.P.; Fisher, R.; Clode, P.L.; Duckworth, A.; Jones, R. Impacts of turbidity on corals: The relative importance of light limitation and suspended sediments. *Mar. Poll. Bull.* **2017**, *117*, 161–170. [CrossRef]

85. Browne, N.; Braoun, C.; McIlwain, J.; Nagarajan, R.; Zinke, J. Borneo coral reefs subject to high sediment loads show evidence of resilience to various environmental stressors. *Peer J.* **2019**, *7*, e7382. [CrossRef]

86. Houlbrèque, F.; Ferrier-Pagès, C. Heterotrophy in tropical scleractinian corals. *Biol. Rev.* **2009**, *84*, 1–17. [CrossRef]

87. Louis, Y.D.; Bhagooli, R.; Kenkel, C.D.; Baker, A.C.; Dyall, S.D. Gene expression biomarkers of heat stress in scleractinian corals: Promises and limitations. *Compar. Biochem. Physiol. Toxic Pharmacol.* **2017**, *191*, 63–77. [CrossRef]

88. Cziesielski, M.J.; Schmidt-Roach, S.; Aranda, M. The past, present, and future of coral heat stress studies. *Ecol. Evol.* **2019**, *9*, 10055–10066. [CrossRef] [PubMed]

89. Mydlarz, L.D.; Fuess, L.; Mann, W.; Pinzón, J.H.; Gochfeld, D.J. Cnidarian immunity: From genomes to phenomes. In *The Cnidaria, Past, Present and Future*; Springer: Cham, Switzerland, 2016; pp. 441–446.

90. Wright, R.M.; Aglyamova, G.V.; Meyer, E.; Matz, M.V. Gene expression associated with white syndromes in a reef building coral *Acropora hyacinthus*. *BMC Genom.* **2015**, *16*, 371. [CrossRef] [PubMed]

91. Seneca, F.O.; Davtian, D.; Boyer, L.; Czerucka, D. Gene expression kinetics of *Exaiptasia pallida* innate immune response to *Vibrio parahaemolyticus* infection. *BMC Genom.* **2020**, *21*, 768. [CrossRef] [PubMed]

92. Mohamed, A.R.; Cumbo, V.R.; Harii, S.; Shinzato, C.; Chan, C.X.; Ragan, M.A.; Satoh, N.; Ball, E.E.; Miller, D.J. Deciphering the nature of the coral–*Chromera* association. *ISME J.* **2018**, *12*, 776–790. [CrossRef]

93. Rodriguez-Lanetty, M.; Harii, S.; Hoegh-Guldberg, O. Early molecular responses of coral larvae to hyperthermal stress. *Mol. Ecol.* **2009**, *18*, 5101–5114. [CrossRef]

94. Vidal-Dupiol, J.; Adjeroud, M.; Roger, E.; Foure, L.; Duval, D.; Mone, Y.; Mitta, G. Coral bleaching under thermal stress: Putative involvement of host/symbiont recognition mechanisms. *BMC Physiol.* **2009**, *9*, 14. [CrossRef]

95. Pinzón, J.H.; Kamel, B.; Burge, C.A.; Harvell, C.D.; Medina, M.; Weil, E.; Mydlarz, L.D. Whole transcriptome analysis reveals changes in expression of immune-related genes during and after bleaching in a reef-building coral. *R. Soc. Open Sci.* **2015**, *2*, 140214. [CrossRef]

96. Rosenzweig, R.; Nillegoda, N.B.; Mayer, M.P.; Bukau, B. The Hsp70 chaperone network. *Nat. Rev. Mol. Cell. Biol.* **2019**, *20*, 665–680. [CrossRef]

97. Rosic, N.N.; Pernice, M.; Dove, S.; Dunn, S.; Hoegh-Guldberg, O. Gene expression profiles of cytosolic heat shock proteins Hsp70 and Hsp90 from symbiotic dinoflagellates in response to thermal stress: Possible implications for coral bleaching. *Cell Stress Chap.* **2011**, *16*, 69–80. [CrossRef]

98. Seveso, D.; Montano, S.; Strona, G.; Orlandi, I.; Galli, P.; Vai, M. The susceptibility of corals to thermal stress by analyzing Hsp60 expression. *Mar. Environ. Res.* **2014**, *99*, 69–75. [CrossRef]

99. Seveso, D.; Arrigoni, R.; Montano, S.; Maggioni, D.; Orlandi, I.; Berumen, M.L.; Galli, P.; Vai, M. Investigating the heat shock protein response involved in coral bleaching across scleractinian species in the central Red Sea. *Coral Reefs* **2020**, *39*, 85–98. [CrossRef]

100. Zheng, B.; Han, M.; Bernier, M.; Wen, J.K. Nuclear actin and actin-binding proteins in the regulation of transcription and gene expression. *FEBS J.* **2009**, *276*, 2669–2685. [CrossRef] [PubMed]

101. DeSalvo, M.K.; Voolstra, C.R.; Sunagawa, S.; Schwarz, J.A.; Stillman, J.H.; Coffroth, M.A.; Szmant, A.M.; Medina, M. Differential gene expression during thermal stress and bleaching in the Caribbean coral *Montastraea faveolata*. *Mol. Ecol.* **2008**, *17*, 3952–3971. [CrossRef] [PubMed]

102. Kaniewska, P.; Campbell, P.R.; Kline, D.I.; Rodriguez-Lanetty, M.; Miller, D.J.; Dove, S.; Hoegh-Guldberg, O. Major cellular and physiological impacts of ocean acidification on a reef building coral. *PLoS ONE* **2012**, *7*, e34659. [CrossRef]

diversity

MDPI

Article

Host Range of the Coral-Associated Worm Snail *Petaloconchus* sp. (Gastropoda: Vermetidae), a Newly Discovered Cryptogenic Pest Species in the Southern Caribbean

Bert W. Hoeksema [1,2,3,*], Charlotte E. Harper [1,2], Sean J. Langdon-Down [1,2], Roel J. van der Schoot [1,2], Annabel Smith-Moorhouse [1,2], Roselle Spaargaren [1,2] and Rosalie F. Timmerman [1,2]

[1] Taxonomy, Systematics and Geodiversity Group, Naturalis Biodiversity Center, P.O. Box 9517, 2300 RA Leiden, The Netherlands; c.harper@student.rug.nl (C.E.H.); s.langdon-down@student.rug.nl (S.J.L.-D.); roel.vanderschoot@naturalis.nl (R.J.v.d.S.); a.smith-moorhouse@student.rug.nl (A.S.-M.); r.spaargaren@student.rug.nl (R.S.); r.f.timmerman@student.rug.nl (R.F.T.)

[2] Groningen Institute for Evolutionary Life Sciences, University of Groningen, P.O. Box 11103, 9700 CC Groningen, The Netherlands

[3] Institute of Biology Leiden, Leiden University, P.O. Box 9505, 2300 RA Leiden, The Netherlands

[*] Correspondence: bert.hoeksema@naturalis.nl

Citation: Hoeksema, B.W.; Harper, C.E.; Langdon-Down, S.J.; van der Schoot, R.J.; Smith-Moorhouse, A.; Spaargaren, R.; Timmerman, R.F. Host Range of the Coral-Associated Worm Snail *Petaloconchus* sp. (Gastropoda: Vermetidae), a Newly Discovered Cryptogenic Pest Species in the Southern Caribbean. *Diversity* 2022, *14*, 196. https://doi.org/10.3390/d14030196

Academic Editor: Savvas Genitsaris

Received: 18 February 2022
Accepted: 4 March 2022
Published: 7 March 2022

Publisher's Note: MDPI stays neutral with regard to jurisdictional claims in published maps and institutional affiliations.

Abstract: The presence of associated endofauna can have an impact on the health of corals. During fieldwork on the southern Caribbean island of Curaçao in 2021, the presence of an unknown coral-dwelling worm snail was discovered, which appeared to cause damage to its hosts. A study of photo archives revealed that the species was already present during earlier surveys at Curaçao since 2014 and also in the southern Caribbean island of Bonaire in 2019. It was not found in St. Eustatius, an island in the eastern Caribbean, during an expedition in 2015. The vermetid snail was preliminarily identified as *Petaloconchus* sp. Its habitat choice resembles that of *P. keenae*, a West Pacific coral symbiont. The Caribbean species was observed in 21 host coral species, more than reported for any other vermetid. Because *Petaloconchus* sp. is a habitat generalist, it is possible that it was introduced from an area with another host-coral fauna. The unknown vermetid is considered to be cryptogenic until future studies reveal its actual identity and its native range.

Keywords: coral damage; coral reef; host generalist; *Millepora*; scleractinia

1. Introduction

Worm snails of the family Vermetidae are common inhabitants of coral reefs and rocky shores in tropical to warm–temperate marine coastal waters, where they live embedded in dead or live corals or attached to other hard substrata [1]. They have tube-shaped shells, mostly without the regular shell coiling [2,3], which in some species form dense, reef-building aggregations [4–6]. Because of their reef-building capacity in the intertidal or immediate subtidal zone, they play an important role as sea-level and sea-surface temperature indicators in the fossil record [6–8]. The history of Vermetidae systematics is complex, which is partly due to the confusion of their calcareous tubes with those of other organisms, such as serpulid worms [9].

Although there is much literature on coral-associated fauna in the Caribbean, vermetid snails are usually not included [10–13]. Apparently, all host-related information on coral-associated vermetids is from the Indo-Pacific, predominantly involving *Ceraesignum maximum* (G.B. Sowerby I, 1825), previously known as *Dendropoma maxima*, which dwells on scleractinians, blue corals, and fire corals [14–16]. This species is notorious because of its harmful effect on the growth, survival, and photophysiology of host corals [17–20]. There are only a few other coral-vermetid records from the Indo-Pacific, including *Petaloconchus keenae* Hadfield & Kay, 1972 near Hawaii [21], *Thylacodes hadfieldi* (W.C. Kelly, 2007) near Guam [22], and *Thylacodes* spp. off the west coast of India [23].

During a recent survey of the coral-associated fauna of Curaçao, coral-dwelling worm snails were discovered for the first time in the Caribbean. In order to investigate their preferred habitats, all observed host coral species were recorded. Because the species appears to be cryptogenic, we discuss why no earlier records are known for the Caribbean. The present report serves to create awareness for this cryptogenic species for future research on its origin and its possible effect on the health of Caribbean coral reefs.

2. Materials and Methods

The survey of coral-associated fauna took place during October–December 2021 along the leeward side of the island of Curaçao. Because the coral-dwelling vermetid was not recorded before in the Caribbean and was overlooked by the first author during earlier surveys, his photo archive was checked for the presence of this snail during fieldwork at Curaçao in 2017, 2015, and 2014, Bonaire in 2019, and St. Eustatius in 2015. Curaçao and Bonaire are located in the southern Caribbean and St. Eustatius in the eastern Caribbean (Figure 1). All association records were listed per island and year (Table 1). Photographic evidence (showing a shell with the operculum present) is presented for each host (Supplementary Materials).

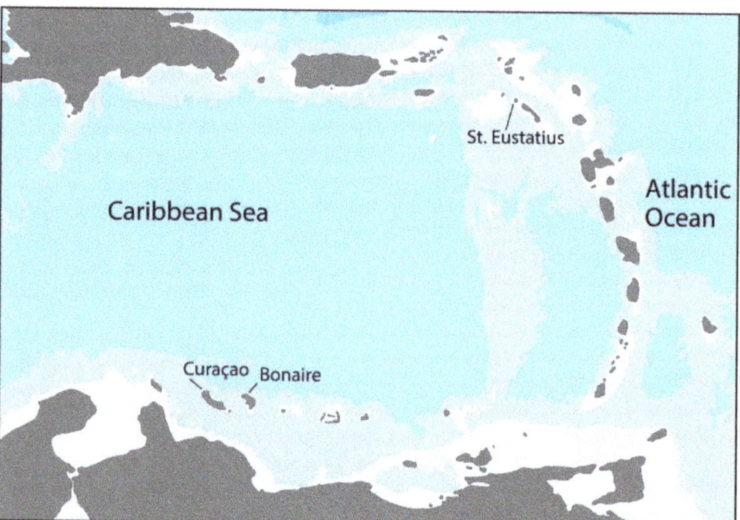

Figure 1. Map of the eastern part of the Caribbean showing the position of Curaçao, Bonaire and St. Eustatius, where the occurrence of coral-dwelling vermetids was investigated.

3. Results

A total of 21 host-coral species—19 scleractinians (Anthozoa) and two milleporids (Hydrozoa)—were recorded, divided over 11 families and 14 genera (Table 1). A few worm snail specimens were found on dead unidentified coral. All records were from the southern Caribbean islands Bonaire and Curaçao, with none from St. Eustatius in the eastern Caribbean. The worm snails with a clearly visible operculum were identified as *Petaloconchus* sp. (Figures 2 and 3; Supplementary Materials Figures S1–S23). Its operculum has upward-folded margins, giving it a tapering appearance, and its diameter was smaller than that of the shells' aperture, which prevented a total shutting of the tube (Figure 3). A small, concave operculum is characteristic for the genus *Petaloconchus* H.C. Lea, 1843 [24].

Table 1. Records of stony corals as host species (by family) for the vermetid gastropod *Petaloconchus* sp. based on photographs made at Curaçao (a: 2021; b: 2017; c: 2015; d: 2014), Bonaire (e: 2019), and St. Eustatius (2015).

Host Species	Curaçao	Bonaire	St. Eustatius
Anthozoa: Scleractinia			
Agariciidae			
Agaricia agaricites (Linnaeus, 1758)	a	e	–
Agaricia humilis (Verrill, 1901)	a	–	–
Agaricia lamarcki Milne Edwards & Haime, 1851	a	–	–
Astrocoeniidae			
Stephanocoenia intersepta (Esper, 1795)	a	–	–
Dendrophylliidae			
Cladopsammia manuelensis (Chevalier, 1966)	b	–	–
Faviidae			
Colpophyllia natans (Houttuyn, 1772)	a	–	–
Diploria labyrinthiformis (Linnaeus, 1758)	a	–	–
Pseudodiploria strigosa (Dana, 1846)	a,d	–	–
Meandrinidae			
Eusmilia fastigiata (Pallas, 1766)	d	–	–
Meandrina meandrites (Linnaeus, 1758)	a	–	–
Merulinidae			
Orbicella annularis (Ellis & Solander, 1786)	a	e	–
Orbicella faveolata (Ellis & Solander, 1786)	a	–	–
Orbicella franksi (Gregory, 1895)	a	e	–
Montastraeidae			
Montastraea cavernosa (Linnaeus, 1767)	a	e	–
Pocilloporidae			
Madracis auretenra Locke, Weil & Coates, 2007	a	–	–
Madracis decactis (Lyman, 1859)	d	e	–
Madracis senaria Wells, 1973	a	e	–
Poritidae			
Porites astreoides Lamarck, 1816	a,b	e	–
Rhizangiidae			
Siderastrea siderea (Ellis & Solander, 1768)	a	–	–
Hydrozoa			
Milleporidae			
Millepora alcicornis Linnaeus, 1758	a	–	–
Millepora complanata Lamarck, 1816	a,c	–	–
Unidentified dead coral	a	–	–

Owing to its symbiotic nature, the coral-dwelling worm snail of the present study cannot be confused with previously reported Caribbean species, such as the reef-building *Petaloconchus varians* (d'Orbigny, 1839) in Venezuela [5] and Brazil [25]. The invasive vermetid *Eualetes tulipa* (Chenu, 1843) has also been recorded in the West Atlantic and the Caribbean, but its operculum is much darker than that of our specimens and it has not been reported as a coral symbiont but as colonies on rock and artificial substrate [26,27]. *Dendropoma corrodens* (d'Orbigny, 1841) is a small worm snail species (ca. 1 cm long), known from the Caribbean and the mid-Atlantic, which forms aggregations on dead coral substrate [26].

Several snail tubes were covered by algae and surrounded by faecal pellets (Figure 3). Many were surrounded by dead coral tissue (Figure 3B,C,E) or attached to dead coral next to the host's margin (Figure 3A). A few snails showed remnants of mucus webs (Figures 2A and 3D,F). In some corals, the snail tube was killing the polyps underneath and did not become overgrown by coral tissue (Figure 2C,D; Supplementary Materials Figures S17 and S19). In other ones no damage was observed, such as in *Cladopsammia manuelensis*, *Eusmilia fastigiata*, *Madracis auretenra*, *M. decactis* and *Millepora alcicornis* (Figure 2A,B; Supplementary

Materials Figures S4 and S7–S9). *Cladopsammia manuelensis* has recently been discovered as a shallow-water coral in the Caribbean [28,29].

Figure 2. Worm snails (*Petaloconchus* sp.) hosted by the branching corals *Madracis auretenra* (**A**) and *Millepora alcicornis* (**B**), and the massive corals *Porites astreoides* (**C**) and *Siderastrea siderea* (**D**). The snail tube in *M. alcicornis* is entirely overgrown by the coral (**B**), whereas the snails in both massive corals have caused considerable damage to their hosts (**C,D**). Tube diameter: ca. 4 mm.

4. Discussion

Since Caribbean coral-dwelling vermetids previously were not recognized in the scientific literature, they may have become introduced recently or they may have been overlooked. The tubes of the snails can be confused with those of polychaete worms of the serpulid genus *Spirobranchus* Blainville, 1818, which are common in the Caribbean, where they have a wide host range [30]. Both groups, coral-dwelling worm snails and serpulid worms, have their tubes partially embedded in the coral skeleton and both possess an operculum that is used to close the tube for the protection of soft body-parts [2,31]. *Spirobranchus* worms are eye-catching because of their high densities and colorful, twin-conispiral branchiae [31,32]. Vermetid snails, on the other hand, use transparent mucus nets to catch food (Figure 3D,F) [33–35], which makes them less remarkable (Supplementary Materials Figure S19).

Figure 3. Close-up images of worm snails (*Petaloconchus* sp.) and their hosts: *Siderastrea siderea* (**A**), *Porites astreoides* (**B,D–F**), and *Madracis auretenra* (**C**). The tubes are partly or entirely overgrown or surrounded by green and red algae (a). The operculum (o) appears to be tapering towards one side because of the margins being turned upwards next to the tentacles. Some worms show remnants of mucus nets (m). Faecal pellets are common around the tubes (fp). Tube diameter: ca. 4 mm.

The habitat of *Petaloconchus* sp. resembles that of *P. keenae*, which has been reported as an associate of the Indo-Pacific coral genera *Porites*, *Montipora*, and *Pavona* at the Hawaiian islands [21], and possibly as *Petaloconchus* cf. *keenae* living in corals at Kwajalein Atoll in the Marshall Islands [36]. It is therefore reasonable to speculate that the coral-associated *Petaloconchus* sp. in the southern Caribbean is the same species and that it has been introduced from the tropical Indo-Pacific. As long as its identity cannot be confirmed by molecular analyses and morphological studies of the radula, protoconch, egg capsules, and coloration of shell and body [21,22], we consider the present species to be cryptogenic in the southern Caribbean. An earlier presence of such Caribbean worm snails could be verified with the help of coral collections in natural history museums [37]. Most museum collections of stony corals consist of dry specimens, but remnants of vermetid shells may still be present and recognizable.

Considering the poor knowledge of *Petaloconchus* sp. in the Caribbean, it is relevant to know of possible natural enemies that may be able to remove evidence of coral-dwelling vermetids. The carpiliid crab *Carpilius convexus* (Forskål, 1775) has been reported to prey on the vermetid *Ceraesignum maximum* in the Red Sea by breaking its shell and the coral in which it lives [38]. A possible predator of *Petaloconchus* sp. would therefore be *Carpilius corallinus* Herbst, 1783, which has been observed to crush large tubes of serpulid worms and also parts of the host coral [39].

The host-coral range in the present study (21 species) is more extensive than recorded for any other coral-associated vermetid. The host ranges of the Indo-Pacific vermetids *C.*

maximum and *P. keenae* (mentioned above) are based on miscellaneous records and might be much larger in reality. Previous field surveys specifically targeting host-coral ranges (and prey preferences) of gastropod families and genera also yielded various additional host records and showed that some species are very host-specific, e.g., *Leptoconchus* [40] and Epitoniidae [41], and that others are generalists, such as some species of *Coralliophila* [42] and *Drupella* [43,44]. Research on coral-dwelling nudibranchs, such as the well-camouflaged *Phestilla* spp., also demonstrates that an ongoing search for possible hosts results in new association records and species discoveries [45–47].

The occurrence of coral injuries (dead coral surface and shells overgrowing coral polyps) suggests that the snails are harmful to their hosts, which is relevant for coral reef conservation [48,49]. It is possible that coral polyps are killed by the snail's mucus webs (Figures 2A and 3D,F), which may smother and even poison them if the mucus is toxic, as reported for *Ceraesignum maximum* [50]. The occurrence of turf algae on the shells is expected to increase damage to the hosts, as observed in *Spirobranchus* tubes overgrowing coral polyps [31,32]. Future field research with a focus on reef-dwelling vermetids will likely result in more information on their densities and additional host records. Molecular analyses may reveal the actual identity of the present species and its native range, and also whether cryptic speciation has taken place across the various host corals.

Supplementary Materials: The following supporting information can be downloaded at: https://www.mdpi.com/article/10.3390/d14030196/s1, Supplementary S1: photographic host records of *Petaloconchus* sp.: Figure S1: *Agaricia agaricites* at Bonaire (2019); Figure S2: *Agaricia humilis* at Curaçao (2021); Figure S3: *Agaricia lamarcki* at Curaçao (2021); Figure S4: *Cladopsammia manuelensis* at Curaçao (2017); Figure S5: *Colpophyllia natans* at Curaçao (2021); Figure S6: *Diploria labyrinthiformis* at Curaçao (2021); Figure S7: *Eusmilia fastigiata* at Curaçao (2014); Figure S8: *Madracis auretenra* at Curaçao (2021); Figure S9: *Madracis decactis* at Bonaire (2019); Figure S10: *Madracis senaria* at Curaçao (2021); Figure S11: *Madracis senaria* at Bonaire (2019); Figure S12: *Meandrina meandrites* at Curaçao (2021); Figure S13: *Millepora alcicornis* at Curaçao (2021); Figure S14: *Millepora complanata* at Curaçao (2015); Figure S15: *Montastraea cavernosa* at Bonaire (2019); Figure S16: *Orbicella annularis* at Bonaire (2019); Figure S17: *Orbicella faveolata* at Curaçao (2021); Figure S18: *Orbicella franksi* at Bonaire (2019); Figure S19: *Porites astreoides* at Curaçao (2021); Figure S20: *Pseudodiploria strigosa* at Curaçao (2021); Figure S21: *Siderastrea siderea* at Curaçao (2021); Figure S22: *Stephanocoenia intersepta* at Curaçao (2021); Figure S23: Unidentified dead coral at Curaçao (2021).

Author Contributions: Conceptualization, B.W.H. and R.J.v.d.S.; methodology, B.W.H., C.E.H., S.J.L.-D., R.J.v.d.S., A.S.-M., R.S., R.F.T.; validation, B.W.H.; formal analysis, B.W.H.; investigation, B.W.H., C.E.H., S.J.L.-D., R.J.v.d.S., A.S.-M., R.S., R.F.T.; resources, B.W.H., C.E.H., S.J.L.-D., R.J.v.d.S., A.S.-M., R.S., R.F.T.; data curation, B.W.H., C.E.H., S.J.L.-D., R.J.v.d.S., A.S.-M., R.S., R.F.T.; writing—original draft preparation, B.W.H.; writing—review and editing, B.W.H., C.E.H., S.J.L.-D., R.J.v.d.S., A.S.-M., R.S., R.F.T.; visualization, B.W.H., C.E.H., S.J.L.-D., R.J.v.d.S., A.S.-M., R.S., R.F.T.; supervision, B.W.H.; project administration, B.W.H.; funding acquisition, B.W.H., C.E.H., S.J.L.-D., R.J.v.d.S., A.S.-M., R.S., R.F.T. All authors have read and agreed to the published version of the manuscript.

Funding: The field research at Curaçao was funded by the Alida M. Buitendijk Fund, the Jan-Joost ter Pelkwijk Fund, the Holthuis Fund, and the Dutch Research Council (NWO) Doctoral Grant for Teachers Programme (nr. 023.015.036). Fieldwork at Bonaire was supported by the World Wildlife Fund (WWF) Netherlands. The Treub Maatschappij (Society for the Advancement of Research in the Tropics) funded research at both Bonaire and Curaçao.

Institutional Review Board Statement: Not applicable.

Data Availability Statement: Data sharing not applicable.

Acknowledgments: We are grateful to the funding agencies mentioned above. We thank Rudiger Bieler (Field Museum of Natural History, Chicago) for the identification of the worm snail. We thank the staff of CARMABI (Curaçao) and the Dive Shop for their hospitality and assistance during the fieldwork. BWH is also grateful to Stichting Nationale Parken Bonaire (STINAPA), Dutch Caribbean Nature Alliance (DCNA) and Dive Friends (Bonaire) for logistical support at Bonaire, and to the Caribbean Netherlands Science Institute (CNSI), St. Eustatius National Parks Foundation (STENAPA), and Scubaqua Dive Center, for facilitating research in St. Eustatius. We also want to thank two anonymous reviewers for their constructive comments, which helped us to improve the manuscript.

Conflicts of Interest: The authors declare no conflict of interest.

References

1. Schiaparelli, S.; Albertelli, G.; Cattaneo-Vietti, R. Phenotypic plasticity of Vermetidae suspension feeding: A potential bias in their use as biological sea-level indicators. *Mar. Ecol.* **2006**, *27*, 44–53. [CrossRef]
2. Morton, J.E. The evolution of vermetid gastropods. *Pac. Sci.* **1955**, *9*, 3–15.
3. Schiaparelli, S.; Cattaneo-Vietti, R. Functional morphology of the vermetid feeding tubes. *Lethaia* **1999**, *32*, 41–46. [CrossRef]
4. Jones, B.; Hunter, I.G. Vermetid buildups from Grand Cayman, British West Indies. *J. Coast. Res.* **1995**, *11*, 973–983. [CrossRef]
5. Weinberger, V.P.; Miloslavich, P.; Machordom, A. Distribution pattern, reproductive traits, and molecular analysis of two coexisting vermetid gastropods of the genus *Petaloconchus*: A Caribbean endemic and a potential invasive species. *Mar. Biol.* **2010**, *157*, 1625–1639. [CrossRef]
6. Montagna, P.; Silenzi, S.; Devoti, S.; Mazzoli, C.; McCulloch, M.; Scicchitano, G.; Taviani, M. Climate reconstructions and monitoring in the Mediterranean Sea: A review on some recently discovered high-resolution marine archives. *Rend. Fis. Acc. Lincei* **2008**, *19*, 121–140. [CrossRef]
7. Vescogni, A.; Bosellini, F.R.; Reuter, M.; Brachert, T.C. Vermetid reefs and their use as palaeobathymetric markers: New insights from the Late Miocene of the Mediterranean (Southern Italy, Crete). *Palaeogeogr. Paleoclimatol. Paleoecol.* **2008**, *267*, 89–101. [CrossRef]
8. Furlani, S.; Vaccher, V.; Antonioli, F.; Agate, M.; Biolchi, S.; Boccali, C.; Busetti, A.; Caldareri, F.; Canziani, F.; Chemello, R.; et al. Preservation of modern and MIS 5.5 erosional landforms and biological structures as sea level markers: A matter of luck? *Water* **2021**, *13*, 2127. [CrossRef]
9. Bieler, R. Mörch's worm-snail taxa (Caenogastropoda: Vermetidae, Siliquariidae, Turritellidae). *Am. Malacol. Bull.* **1996**, *13*, 23–35.
10. Scott, P.J.B. Associations between corals and macro-infaunal invertebrates in Jamaica, with a list of Caribbean and Atlantic coral associates. *Bull. Mar. Sci.* **1987**, *40*, 271–286.
11. Perry, C.T. Macroborers within coral framework at Discovery Bay, north Jamaica: Species distribution and abundance, and effects on coral preservation. *Coral Reefs* **1998**, *17*, 277–287. [CrossRef]
12. Lewis, J.B. Biology and ecology of the hydrocoral *Millepora* on coral reefs. *Adv. Mar. Biol.* **2000**, *50*, 1–55. [CrossRef]
13. Hoeksema, B.W.; van Beusekom, M.; ten Hove, H.A.; Ivanenko, V.N.; van der Meij, S.E.T.; van Moorsel, G.W.N.M. *Helioseris cucullata* as a host coral at St. Eustatius. Dutch Caribbean. *Mar. Biodivers.* **2017**, *47*, 71–78. [CrossRef]
14. Colgan, M.W. Growth rate reduction and modification of a coral colony by a vermetid mollusc *Dendropoma maxima*. In Proceedings of the 5th International Coral Reef Congress, Tahiti, 27 May–1 June 1985; Volume 6, pp. 205–210.
15. Scaps, P.; Denis, V. Can organisms associated with live scleractinian corals be used as indicators of coral reef status? *Atoll Res. Bull.* **2008**, *566*, 1–18. [CrossRef]
16. Zvuloni, A.; Armoza-Zvuloni, R.; Loya, Y. Structural deformation of branching corals associated with the vermetid gastropod *Dendropoma maxima*. *Mar. Ecol. Prog. Ser.* **2008**, *363*, 103–108. [CrossRef]
17. Shima, J.S.; Osenberg, C.W.; Stier, A. The vermetid gastropod *Dendropoma maximum* reduces coral growth and survival. *Biol. Lett.* **2010**, *6*, 815–818. [CrossRef]
18. Shima, J.S.; Phillips, N.E.; Osenberg, C.W. Consistent deleterious effects of vermetid gastropods on coral performance. *J. Exp. Mar. Biol. Ecol.* **2013**, *439*, 1–6. [CrossRef]
19. Brown, A.L.; Osenberg, C.W. Vermetid gastropods modify physical and chemical conditions above coral–algal interactions. *Oecologia* **2018**, *186*, 1091–1099. [CrossRef]
20. Barton, J.A.; Bourne, D.G.; Humphrey, C.M.; Hutson, K.S. Parasites and coral-associated invertebrates that impact coral health. *Rev. Aquac.* **2020**, *12*, 2284–2303. [CrossRef]
21. Hadfield, M.G.; Kay, E.A.; Gillette, M.U.; Lloyd, M.C. The Vermetidae (Mollusca: Gastropoda) of the Hawaiian Islands. *Mar. Biol.* **1972**, *12*, 81–98. [CrossRef]
22. Kelly, W.C., III. Three new vermetid gastropod species from Guam. *Micronesica* **2007**, *39*, 117–140.
23. Joshi, D.M.; Mankodi, P.C. The Vermetidae of the Gulf of Kachchh, western coast of India (Mollusca, Gastropoda). *ZooKeys* **2016**, *555*, 1–10. [CrossRef]
24. Barash, A.; Zenziper, Z. Structural and biological adaptations of Vermetidae (Gastropoda). *Boll. Malacol.* **1985**, *21*, 145–176.
25. Breves, A.; Széchy, M.T.M.; Lavrado, H.P.; Junqueira, A.O. Abundance of the reef-building *Petaloconchus varians* (Gastropoda: Vermetidae) on intertidal rocky shores at Ilha Grande Bay, southeastern Brazil. *An. Acad. Bras. Ciências* **2017**, *89*, 907–918. [CrossRef]
26. Miloslavich, P.; Klein, E.; Penchaszadeh, P. Gametogenic cycle of the tropical vermetids *Eualetes tulipa* and *Dendropoma corrodens* (Mollusca: Caenogastropoda: Vermetidae). *J. Mar. Biol. Assoc. UK* **2010**, *90*, 509–518. [CrossRef]
27. Spotorno-Oliveira, P.; Coutinho, R.; de Souza Tâmeg, F.T. Recent introduction of non-indigenous vermetid species (Mollusca, Vermetidae) to the Brazilian coast. *Mar. Biodivers.* **2018**, *48*, 1931–1941. [CrossRef]

28. Hoeksema, B.W.; Hiemstra, A.F.; Vermeij, M.J.A. The rise of a native sun coral species on southern Caribbean coral reefs. *Ecosphere* **2019**, *10*, e02942. [CrossRef]

29. Hammerman, N.M.; Williams, S.M.; Veglia, A.J.; Garcia-Hernandez, J.E.; Schizas, N.V.; Lang, J.C. *Cladopsammia manuelensis* sensu Cairns, 2000 (Order: Scleractinia): A new distribution record for Hispaniola and Puerto Rico. *Cah. Biol. Mar.* **2021**, *62*, 1–10. [CrossRef]

30. Hoeksema, B.W.; ten Hove, H.A. The invasive sun coral *Tubastraea coccinea* hosting a native Christmas tree worm at Curaçao, Dutch Caribbean. *Mar. Biodivers.* **2017**, *47*, 59–65. [CrossRef]

31. Hoeksema, B.W.; Wels, D.; van der Schoot, R.J.; ten Hove, H.A. Coral injuries caused by *Spirobranchus* opercula with and without epibiotic turf algae at Curaçao. *Mar. Biol.* **2019**, *166*, 60. [CrossRef]

32. Hoeksema, B.W.; van der Schoot, R.J.; Wels, D.; Scott, C.; ten Hove, H.A. Filamentous turf algae on tube worms intensify damage in massive *Porites* corals. *Ecology* **2019**, *100*, e2668. [CrossRef] [PubMed]

33. Hughes, R.N.; Lewis, A. On the spatial distribution, feeding and reproduction of the vermetid gastropod *Dendropoma maximum*. *J. Zool.* **1974**, *172*, 531–547. [CrossRef]

34. Kappner, I.; Al-Moghrabi, S.M.; Richter, C. Mucus-net feeding by the vermetid gastropod *Dendropoma maxima* in coral reefs. *Mar. Ecol. Prog. Ser.* **2000**, *204*, 309–313. [CrossRef]

35. Kusama, Y.; Nakano, T.; Asakura, A. Mucus-net feeding behavior by the sessile gastropod *Thylacodes adamsii* (Gastropoda: Vermetidae). *Publ. Seto Mar. Biol. Lab.* **2021**, *46*, 55–69. [CrossRef]

36. Kwajalein Underwater: *Petaloconchus* cf *keenae* Hadfield & Kay. 1972. Available online: http://www.underwaterkwaj.com/shell/vermetid/Petaloconchus-cf-keenae.htm (accessed on 14 February 2022).

37. Hoeksema, B.W.; van der Land, J.; van der Meij, S.E.T.; van Ofwegen, L.P.; Reijnen, B.T.; van Soest, R.W.M.; de Voogd, N.J. Unforeseen importance of historical collections as baselines to determine biotic change of coral reefs: The Saba Bank case. *Mar. Ecol.* **2011**, *32*, 135–141. [CrossRef]

38. Shlesinger, T.; Akkaynak, D.; Loya, Y. Who is smashing the reef at night? A nocturnal mystery. *Ecology* **2021**, *102*, e03420. [CrossRef]

39. Muller, E.; de Gier, W.; ten Hove, H.A.; van Moorsel, G.W.N.M.; Hoeksema, B.W. Nocturnal predation of Christmas tree worms by a Batwing coral crab at Bonaire (Southern Caribbean). *Diversity* **2020**, *12*, 455. [CrossRef]

40. Gittenberger, A.; Gittenberger, E. Cryptic, adaptive radiation of endoparasitic snails: Sibling species of *Leptoconchus* (Gastropoda: Coralliophilidae) in corals. *Org. Divers. Evol.* **2011**, *11*, 21–41. [CrossRef]

41. Gittenberger, A.; Hoeksema, B.W. Habitat preferences of coral-associated wentletrap snails (Gastropoda: Epitoniidae). *Contrib. Zool.* **2013**, *82*, 1–25. [CrossRef]

42. Potkamp, G.; Vermeij, M.J.A.; Hoeksema, B.W. Genetic and morphological variation in corallivorous snails (Coralliophila spp.) living on different host corals at Curaçao, southern Caribbean. *Contrib. Zool.* **2017**, *86*, 111–144. [CrossRef]

43. Hoeksema, B.W.; Scott, C.; True, J.D. Dietary shift in corallivorous *Drupella* snails following a major bleaching event at Koh Tao, Gulf of Thailand. *Coral Reefs* **2013**, *32*, 423–428. [CrossRef]

44. Moerland, M.S.; Scott, C.M.; Hoeksema, B.W. Prey selection of corallivorous muricids at Koh Tao (Gulf of Thailand) four years after a major coral bleaching event. *Contrib. Zool.* **2016**, *85*, 291–309. [CrossRef]

45. Mehrotra, R.; Arnold, S.; Wang, A.; Chavanich, S.; Hoeksema, B.W.; Caballer, M. A new species of coral-feeding nudibranch (Mollusca: Gastropoda) from the Gulf of Thailand. *Mar. Biodivers.* **2020**, *50*, 36. [CrossRef]

46. Wang, A.; Conti-Jerpe, I.E.; Richards, J.L.; Baker, D.M. *Phestilla subodiosus* sp. nov. (Nudibranchia, Trinchesiidae), a corallivorous pest species in the aquarium trade. *ZooKeys* **2020**, *909*, 1–24. [CrossRef] [PubMed]

47. Yiu, S.K.F.; Chung, S.S.W.; Qiu, J.W. New observations on the corallivorous nudibranch *Phestilla melanobrachia*: Morphology, dietary spectrum and early development. *J. Molluscan Stud.* **2021**, *87*, eyab034. [CrossRef]

48. Stella, J.S.; Pratchett, M.S.; Hutchings, P.A.; Jones, G.P. Coral-associated invertebrates: Diversity, ecology importance and vulnerability to disturbance. *Oceanogr. Mar. Biol. Ann. Rev.* **2011**, *49*, 43–104.

49. Montano, S. The extraordinary importance of coral-associated fauna. *Diversity* **2020**, *12*, 357. [CrossRef]

50. Klöppel, A.; Brümmer, F.; Schwabe, D.; Morlock, G. Detection of bioactive compounds in the mucus nets of *Dendropoma maxima*, Sowerby 1825 (Prosobranch Gastropod Vermetidae, Mollusca). *J. Mar. Biol.* **2013**, *2013*, 283506. [CrossRef]

diversity

MDPI

Article

Re-Examination of the Phylogenetic Relationship among Merulinidae Subclades in Non-Reefal Coral Communities of Northeastern Taiwan

Chieh-Jhen Chen [1,2], Yu-Ying Ho [1] and Ching-Fong Chang [1,2,*]

[1] Department of Aquaculture, National Taiwan Ocean University, Keelung 20224, Taiwan; coralerin2933@gmail.com (C.-J.C.); onlyoneian7@gmail.com (Y.-Y.H.)
[2] Center of Excellence for the Oceans, National Taiwan Ocean University, Keelung 20224, Taiwan
* Correspondence: b0044@email.ntou.edu.tw

Abstract: Species identification for spawning corals relies heavily on morphology. Recent molecular phylogenetic approaches have demonstrated the limits of traditional coral taxonomy based solely on skeletal morphology. Merulinidae is considered a complex taxonomic group, containing 24 genera and 149 species. This family is one of the most taxonomically challenging and its taxonomy has largely improved in recent studies. However, studies of the phylogeny of Merulinidae are constrained by limited geographic scales. In Taiwan, merulinid corals are dominant in non-reefal communities on northeast coasts and they consistently spawn between summer and fall. This study is a first attempt to establish a molecular database of merulinid corals in this new area, including a volcanic island (Kueishan Island), and provide information about sexual reproduction. We analyzed 65 specimens, including 9 genera and 28 species collected from Taiwan using one mitochondrial marker (COI: cytochrome c oxidase subunit 1 gene) and three nuclear markers (ITS: nuclear ribosomal internal transcribed spacer, 28S rDNA D1 and D2, and histone H3) to re-examine phylogenetic relationships and search for new species. Overall, 58 COI sequences, 59 for ITS, 63 for 28S, and 62 histone sequences were newly obtained from the collected specimens. The reconstructed molecular tree demonstrates that all the specimens and reference sequences we examined are clustered within Merulinidae. Subclades A, B, C, D/E, F, G, H, and I are congruent with previous studies. However, *Astrea curta* is separated from the other congeneric species, *Astrea annuligera* (XVII-B), which is a sister to *Favites* and defined as a new subclade K. In addition, two new species (*Paragoniastrea deformis* and *Paragoniastrea australensis*) were discovered for the first time in Taiwan, and we defined them as a new subclade J. In addition, *A. curta*, *P. auastralensis*, and *P. deformis* are all hermaphroditic spawners and released bundles in July. This study greatly improves the accuracy of biodiversity estimates, systematic taxonomy, and reproduction for Taiwan's coral ecosystem.

Keywords: taxonomy; Taiwanese corals; molecular phylogeny; scleractinian corals; reproduction

Citation: Chen, C.-J.; Ho, Y.-Y.; Chang, C.-F. Re-Examination of the Phylogenetic Relationship among Merulinidae Subclades in Non-Reefal Coral Communities of Northeastern Taiwan. *Diversity* **2022**, *14*, 144. https://doi.org/10.3390/d14020144

Academic Editors: Michael Wink and Simone Montano

Received: 29 December 2021
Accepted: 14 February 2022
Published: 17 February 2022

1. Introduction

1.1. Uncovering Taxonomic Progress

The identification of scleractinian corals based solely on morphology is challenging because some scleractinian species can exhibit environment-correlated variations in morphology, i.e., Ecomorphs [1]. In addition, species display phenotypic plasticity across their distribution, making it difficult to rely on shared morphological features to identify them [2,3]. Therefore, it is important to combine morphological and molecular characteristics to improve the accuracy of the determination of evolutionary relationships.

The traditional classification of scleractinia into seven suborders was out of date [4–7]. Given the comprehensive study of the entire taxon with morphological and molecular approaches, the scleractinian corals can be generally divided into three major groups: basal, robust and complex [8]. Furthermore, they are separated into 21 clades (I-XXI) [8,9]. Many scleractinian corals at family and genus were revised or remained unclear taxonomic position (Scleractinia *incertae sedis*). For example, *Diploastrea helipora* and *Montastraea carvernosa* were separated into Diploastraeidae Chevalier & Beauvais, 1987, and Montastraeaidae Yabe & Sugiyama, 1941, respectively [10]. Euphylliidae Milne Edward & Haime, 1857, contains six genera: *Ctenella* Matthai, 1982; *Euphyllia* Dana, 1846; *Galaxea* Oken, 1815; *Gyrosmilia* Milne Edwards & Haime, 1851; *Montigyra* Matthai, 1928; *Simplastrea* Umbgrove, 1939; and *Frimbriaphyllia* Veron & Pichon, 1980, the last of which was redefined from the conventional *Euphyllia ancora*, *E. yaeyamaensis*, and *E. divisa* [11]. In addition, the genera *Nemenzophyllia*, *Physogyra*, and *Plerogyra* were removed from Euphylliidae because they formed a separate clade with *Blastomussa* (clade XIV) [9,12].

1.2. Revision of Merulinidae (Clade XVII)

The species identification of Faviidae, Gregory, 1900 and Wells, 1956 was based on their budding patterns and macromorphological characteristics They were traditionally subdivided into two subfamilies based on whether their budding was primarily intracalicular (*Caulastraea*, *Favia*, *Diploria*, *Favites*, *Oulophyllia*, *Goniastrea*, *Platygyra*, *Leptoria*, *Hydnophora*, *Manicina*, and *Colpophyllia*) or extracalicular (*Montastraea*, *Diploastrea*, *Cyphastrea*, and *Echinopora*). A third, smaller, subfamily displays intracalicular budding and very well-developed septal lobes (trabecular versus lamellar, continuous versus discontinuous). Genera within the Faviinae are distinguished by having a colony form (ceroid versus plocoid, mendroid versus phaceloid) and the columella structure (trabecular versus lamellar versus continuous versus discontinuous).

Based on the molecular results, the genera of Faviidae not only displayed a polyphyletic pattern, but were also clustered together with species from four conventional coral families: Faviidae Milne Edwards & Haime, 1857; Merulinidae Verrill, 1995; Pectinidae Rafinesque, 1815; and Trachyphylliidae Well, 1956, which had previously been recovered as Merulinidae (XVII) [10,13–17]. The faviid corals outside of clade XVII were assigned to other families. For example: *Plesiastrea versipora*, *Diploastrea helipora*, and *Montastraea cavernosa* were reclassified as Plesiastreidae Dai & Horng, 2009, Diploastraeidae Chevalier & Beauvais, 1987, and Montastraeidae Yabe & Sugiyama, 1941, respectively. Faviidae is limited to Atlantic corals such as *Favia*, *Diploria*, and *Manicina* because they are evolutionarily divergent to the Pacific corals [18,19]. Furthermore, phylogenies based on multiple genetic markers and morphological characteristics demonstrated that the species/genera Merulinidae are divided into nine subclades (A, B, C, D/E, F, G, H, and I) [10,14,16]. *Paramontastraea*, *Orbicella*, and *Astrea* are new genera in the Merulinidae, revised from *Montastraea*. Given these results, Merulinidae contains the most genera (with 24) and the second-most species (with 149) among the scleractinians [20] (Supplementary Table S1). Its species are commonly distributed in the Indo-Pacific [2,21].

1.3. Taiwan Taxonomy and Species Diversity

Taiwan is located at the center of the Philippine–Japan Island arc at a latitude of 21.90° N to 25.3° N, crossing from the Tropic of Cancer close to the northern tip of the Coral Triangle [22]. To date, 317 scleractinian coral species have been reported in Taiwan and display a latitudinal gradient of decreasing species diversity from south to north [23,24]. In addition, coral assemblages contain 21 genera covering 87% of the total number of genera of merulinid corals and 89 species covering 60% of the total number of merulinid species [25]. Taxonomic phylogenetic studies of scleractinian corals collected from Taiwan are very limited [9,11,26–29]. In addition, biogeographical integration is needed on a larger scale. For example, *Polycyathus chaishanensis* (Caryophyllidae) was proposed to be endemic to Taiwan [27]. Later, this species was also found to inhabit Indonesia, based on molecular evidence [30]. *Euphyllia ancora* has been a model species for studies on sexual reproduction [31] and its genus was recently revised to *Fimbraphyllia* [11]. This revision created an important foundation on the convergent and divergent functionalities of genes and compared functional genes among the cnidarians underlaying precisely the phylogenetic position of the studied species.

1.4. Purpose of This Research

The phylogeny of Merulinidae reconstructed in Huang et al. [16] was based on samples/taxa from Australia, Singapore, Japan, and the Philipines in the Pacific Ocean and the Atlantic Ocean. Taiwan is located in the Pacific Ocean; it is an important stepping stone between the Philippines and Japan. Merulinid corals are major spawning members and consistently spawn between summer and fall in non-reefal coral communities in northern Taiwan [32]. These spawning corals are important for maintaining local recruitment, providing heterogenetic materials to the local population and connecting across different populations. However, some convergent macro-morphological characteristics make it challenging to identify some genera in the field, such as *Goniastrea* (ceroid form), *Favites* (ceroid and plocoid forms), and *Diploastraea* (plocoid form) [24,33]. In addition, species identification for spawning corals based on morphological criteria in the fields and underwater photographs is difficult because the polyps are deformed when "the mature sperm and eggs move to the mouths of polys" (i.e., bundle setting).

As mentioned above, these challenges can be resolved by molecular approaches, as was demonstrated by Huang et al. [10]. For example, ceroid forms of *Goniastrea*, *Diploastraea*, and *Favites* are clearly separated in subclades A, B, and F based on phylogenetic reconstruction using multiple loci [10]. Therefore, Chen et al. [32] identified the species to the genus level of each specimen using molecular approaches and the BLAST tool [34]. Subsequently, specimens were identified to species level using the morphology of their skeletons. The established molecular database of merulinid corals can provide further insight into the phylogenetic relationships among the subclades of Merulinidae. The objectives in this present study were to: (1) establish a molecular database of spawning corals of Merulinidae from Taiwan, which have not been studied before; (2) re-examine the phylogenetic relationship between the specimens collected from northern Taiwan and the merulinid corals in previous studies, using phylogeny reconstructions based on multiple loci; and (3) record any new species or subclades we might find in this region.

2. Materials and Methods

2.1. Sample Collection

Chen et al. [32] demonstrated that the spawning season for merulinid corals is July to August, from 2014 to 2016, in northeast Taiwan. Merulinid corals with bundle-setting behavior and released bundles still attached outside of the mouths of polyps were collected at night by scuba diving. Some corals were collected at two offshore islands and their sexual reproductive behavior was observed using histological approaches [32]. A total of 65 specimens from four sites were chosen for this study: 26 specimens from Pitoujiiao (25°07′34″ N, 121°54′55″ E), 21 from Longdong (25°05′02″ N, 121°55′09″ E), 3 from Keelung Island (25°07′34″ N, 121°54′55″ E), and 12 from Kueishan Island (24°84′19″ N, 121°57′06″ E) (Figure 1). Coral fragments were collected by using chisels and hammers and separated into two parts. One was fixed in 90% ethanol for molecular analysis. The other was bleached in sodium hypochlorite until the tissue was entirely removed, rinsed in freshwater, and air-dried for the morphological analysis.

2.2. Species Identification

Chen et al. [32] identified 54 coral species in 23 genera and 8 families (Acroporidae, Agariciidae, Fungiidae, Lobophylliidae, Merulinidae, Poritidae, Pocilloporidae, and Psammocoridae), which were sexually reproductive between July and October. For Merulinidae, nine genera and 26 species collected from northeast Taiwan were chosen for the molecular phylogenetic study: *Astrea curta* (*n* = 5), *Astrea annuligera* (*n* = 1), *Coelastrea aspera* (*n* = 2), *Coelastrea palauensis* (*n* = 1), *Cyphastrea chalcidicum* (*n* = 2), *Dipsastraea favus* (*n* = 4), *Dipsastraea lizardensis* (*n* = 1), *Dipsastraea matthaii* (*n* = 1), *Dipsastraea rotumana* (*n* = 1),

Favites flexuosa (*n* = 1), *Favites pentagona* (*n* = 7), *Favites stylifera* (*n* = 2), *Favites magnistellata* (*n* = 2), *Favites valenciennesi* (*n* = 2), *Mycedium elephantotus* (*n* = 1), *Mycedium robokaki* (*n* = 1), *Mycedium mancaoi* (*n* = 1), *Paragoniastraea australensis* (*n* = 5), *Paragoniastrea deformis* (*n* = 6), *Pectinia paeonia* (*n* = 1), *Pectinia lactuca* (*n* = 1), *Platygyra daedalea* (*n* = 1), *Platygyra lamellina* (*n* = 2), *Platygyra ryukyuensis* (*n* = 5), *Platygyra pini* (*n* = 2), *Platygyra sinensis* (*n* = 1), and *Platygyra verweyi* (*n* = 3). Those specimens were identified to the genus level using molecular sequences and BLAST searches (http://www.ncbi.nlm.nih.gov/BLAST/, accessed on 2 April 2020) [34]. Subsequently, individuals were identified to the species level using morphological keys, notably Dai and Cheng [25]. Specimens that could not be identified morphologically (cerioid corals: *Goniastrea* and *Favites*, plocoid corals: *Favites* and *Dipsastraea*, unknown species, etc.) were preliminarily identified to the genus level and then re-evaluated after molecular analyses. DNA extraction, PCR amplification, and sequencing

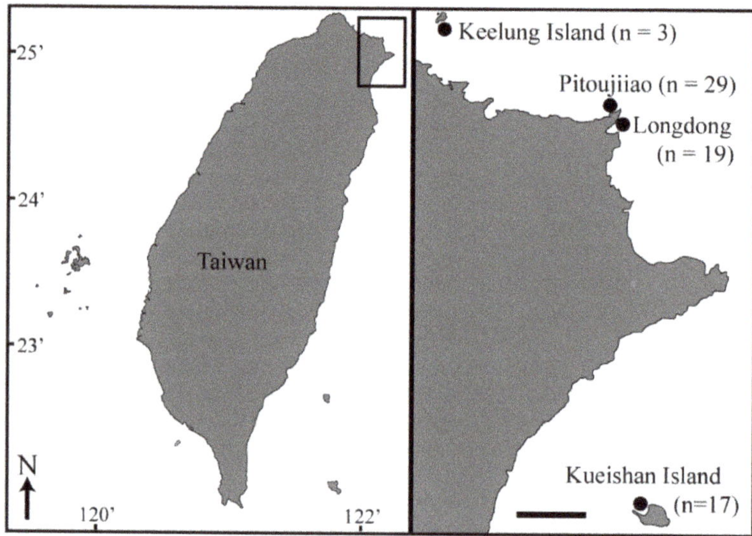

Figure 1. Map showing sampling sites at Pitoujiiao, Longdong, Keelung Island, and Kueishan Island in northeastern Taiwan.

Genomic DNA was extracted from 90% ethanol-preserved tissue specimens using the automated LabTurbo Nucleic Acid Mini Kit LGD480-220 (Taigen Bioscience Corporation), following the manufacturer's protocols. A total of four genes were amplified from the collected specimens, including one mitochondrial marker and three nuclear markers, following Huang et al. [16]: (1) cytochrome c oxidase subunit I segment (MCOIF: 5′-TCTACAAATCATAAAGACATAGG-3′, MCOIR:5′-GAGAAATTATACCAAAACCAGG-3′); (2) nuclear ribosomal internal transcribed spacer segment (ITS, A18S: 5′-GATCGAACGGTTT AGTGAGG-3′, ITS-4: 5′-TCCTCCGCTTATTGATATGC-3′); (3) two variable domain (D1 and D2) at 5′end of 28S ribosomal RNA segment (C1′: 5′-ACCCGCTGAATTTAAGCAT-3′, D2MAD: 5′-GACGATCGATTTGCACGTCA-3′); and (4) histone H3 segment (H3F: 5′-ATGGCTCGTACCAAGCAGACVGC-3′, H3R: 5′-ATATCCTTR GGCATRATRGTGAC-3′). PCR was carried out using 12.5 µL of Fast-Run^{TM} Advanced Taq Master Mix (Protech, Taipei, Taiwan), 10 mM each of forward and reverse ITS primer, 10–100 ng/µL DNA template, and deionized water to a final volume of 25 µL. The PCR profiles were as follows: an initial denaturation stage (95 °C, 5 min); 35 cycles of a denaturation step (95 °C, 30 s, an annealing step (54 °C, 40 s); an elongation step (72 °C, 7 min); and a final extension at 72 °C, for 5 min. The PCR products were confirmed by electrophoresis and subcloned into a pGEM-T easy vector (Promega, Madison, WI, USA). Three inserted cDNA fragments were sequenced with the pUC/M13 forward and reverse primers using an ABI Prism 310 Genetic Analyzer (Applied Biosystems, Forster City, CA, USA).

2.3. Sequence Management, Alignment, and Matrix

The raw forward and reverse sequences were edited and assembled into consensus sequences by the CodonCode Aligner V6.0.2 program (CodonCode Corporation Dedham, MA, USA). To exclude sequences amplified from zooxanthellae, the consensus sequences obtained were used to perform the BLAST searches (http://www.ncbi.nlm.nih.gov/BLAST/, accessed on 2 April 2020) [34]. The sequences obtained from the collected specimens of spawning corals in northern Taiwan were deposited into the NCBI GenBank (accession numbers in Supplementary Table S1). Newly obtained sequences for COI ($n = 58$), ITS ($n = 59$), 28S ($n = 63$), and histone ($n = 62$) were combined with sequences retrieved from public sources (Table S1).

All sequences for each gene were automatically aligned with the accurate alignment option (E-INS-i) in MAFFT v.7 ([35]; http://mafft.cbrc.jp/alighment/server/, accessed on 10 January 2021) under default parameters. The resulting multiple sequence alignments were translated into inferred amino acid sequences as a guide for inferred gap placement between coding regions using Se-Al v.2.0a11 [36]. The amino acid residue and nucleotide were manually adjusted to minimize the gaps. PAUPRat software v.3.1 [37] on the CIPRES Science Gateway (http://www.phylo.org, accessed on 10 January 2021) [38] was used to calculate descriptive statistics (sequence variations and informative sites) for the compared sequences of each gene.

2.4. Molecular Datasets

Some sequences were not obtained because the gene failed to amplify during PCR. Operational taxonomic units (OTUs) were created for each gene for the phylogeny reconstruction. The phylogeny reconstructions were conducted based on the combined gene matrix (COI, 28S, ITS, and histone); IGR (noncoding intergenic region between COI and the formylmethionine transfer RNA gene) was ignored because the overall alignment was not similar enough.

The sequences of merulinid corals published in Huang et al. [16] were retrieved from GenBank. They included 19 genera in Merulinidae: *Merulina* (2 species), *Caulastraea* (3 species), *Cyphastrea* (3 species), *Dipsastraea* (13 species), *Echinopora* (5 species), *Favites* (13 species), *Goniastrea* (5 species), *Hydnophora* (2 species), *Leptoria* (2 species), *Mycedium* (2 species), *Orbicella* (1 species), *Oulophyllia* (2 species), *Pectinia* (3 species), *Platygyra* (8 species), *Scapophyllia* (1 species), and *Trachyphyllia* (1 species); two resurrected genera (*Astrea* (2 species), *Coelastrea* (2 species)); and one new genus (*Paramontastraea* (1 species)). In total, we retrieved 124 sequences for 28S rDNA, 121 sequences for histone H3, 91 for ITS rDNA, and 112 for COI from the GenBank.

The clades distant from the merulinid corals (XVIII-XXI) were included for the phylogenetic inference following Huang et al. [10]. These sequences were comprised of three species of Lobophylliidae (*Moseleya latistellata*, *Acanthastrea echinata*, and *Lobophyllia corymbosa*), three species of Faviidae (*Montastraea multipunctata*, *Favia fragum*, and *Mussa angulosa*), and one species of Plesiastreidae (*Plesiastrea versipora*). Phylogeny reconstructions were created for each gene, along with four combined datasets based on maximum likelihood and Bayesian analyses.

2.5. Molecular Phylogenetic Analysis

The maximum likelihood (ML) trees of each partition were reconstructed with raxml-GUI v.2.0 [39] using the best model (GTR+I+G). The five datasets, including three nuclear genes (ITS, 28S, and histone H3), one mitochondrial gene (COI), and a combined gene dataset, were partitioned based on coding position. The combined gene datasets were conducted with five independent runs, and the tree with the best ML scores was selected as the final tree. Nodal support was assessed by bootstrapping, and only the nodes with ≥70 [40] based on 1000 pseudo-replicates were shown.

Bayesian inference (BI) was carried out in MrBayes v.3.2.6 [41]. PartitionFinder was used to select the best partition scheme and accompanying substitution model, according to the Bayesian information criterion [42]. The best-fit substitution model was determined by ProtTest3. Two Monte Carlo Markov chains (MCMCs) were run for 4×10^6 million generations in two simultaneous runs, each with four different chains. The convergence of the estimates was checked by the standard deviation of split frequencies and by monitoring the likelihood score over time using Tracer v.1.6 [43]. Trees were sampled every 1000 generations, with the first 2500 (25%) discarded as "burn-in." The remaining sampled trees were collected to construct a 50% majority-rule BI consensus tree. Nodal support from BI was assessed, and only nodes with ≥0.90 posterior probabilities (PPs) were shown.

The output trees were further edited by FigTree v1.3.1 [44]. *Plesiastrea versipora* (clade XIV, Plesiastreidae) was set as a distant outgroup to root the inferred trees. The subclades within Merulinidae (XVII) were divided into subclades A-I, following Huang et al. [10,16].

3. Results

3.1. Characteristics of the Gene Data

In the 65 specimens collected from Taiwan, 58 newly obtained COI sequences, 59 newly ITS sequences, 63 for 28S, and 62 for histone H3 sequences were obtained for the first time (Table S1). Examining the individual gene dataset, the aligned COI sequence was 744 base pairs (bp) long, with 180 variable and 81 parsimony informative sites. That of 28S was 865 bp, with 316 variable and 147 parsimony informative sites. That of ITS was 1249 bp, with 662 variable and 459 parsimony informative sites. That of histone H3 was 344 bp, with 109 variable and 82 parsimony informative sites.

3.2. Results of the Analysis Matrix

The dataset comprised a total of 3202 bp and 186 OTUs (123 OTUs from references and 65 OTUs from the present study). ML and BI methods (using raxmlGUI and MrBayes, respectively) were used to reconstruct the phylogenies for the combined dataset. The results from the partitioned ML analysis and BI conducted with the combined dataset were congruent (Figure 2). The ML analysis yielded a log-likelihood value of −28,909.928401 and the BI analysis yielded ($−3.069135 \times 10^4$, $−3.083239 \times 10^4$).

3.3. Phylogenetic Relationship

The clades XV (Diploastreidae), XVI (Montasraeidae), and XVII (Merulinidae) were monophyletic, with high ML and BI support (100/1 and 100/0.99), whereas clades XVIII-XX (Lobophylliidae) and XXI (Mussidae) formed clusters but with weak support (63/0.8 and 58/0.84) (Figure 2). Within Merulinidae, eight major subclades (A, C, D/E, F, H, and I) formed with high ML (83–100) and BI (100) support. Subclades B and G, on the other hand, did not have high support (84/− and −/0.99). Subclade A is composed of *Paragoniastrea australensis*, *Scapophyllia cylindrica*, as well as species of *Goniastrea* and *Merulina*. Subclade B is composed of *Astrea annuligera*, *Favites valenciennesi*, and *Trachyphyllia geoffroyi*, as well as species of *Coelastrea* and *Dipsastraea*. Subclade C is composed of *Orbicella annularis* and species of *Cyphastrea*. Subclade D/E is composed of species of *Caulastrea*, *Oulophyllia*, *Mycedium*, and *Pectinia*. Subclade F is composed of species of *Favites*. Subclade G is composed of *Favites stylifera* and species of *Platygyra* and *Leptoria*. Subclade H is composed of species of *Hydnophora*. Finally, subclade I is composed of species of *Echinopora* and *Paramontastraea salebrosa*.

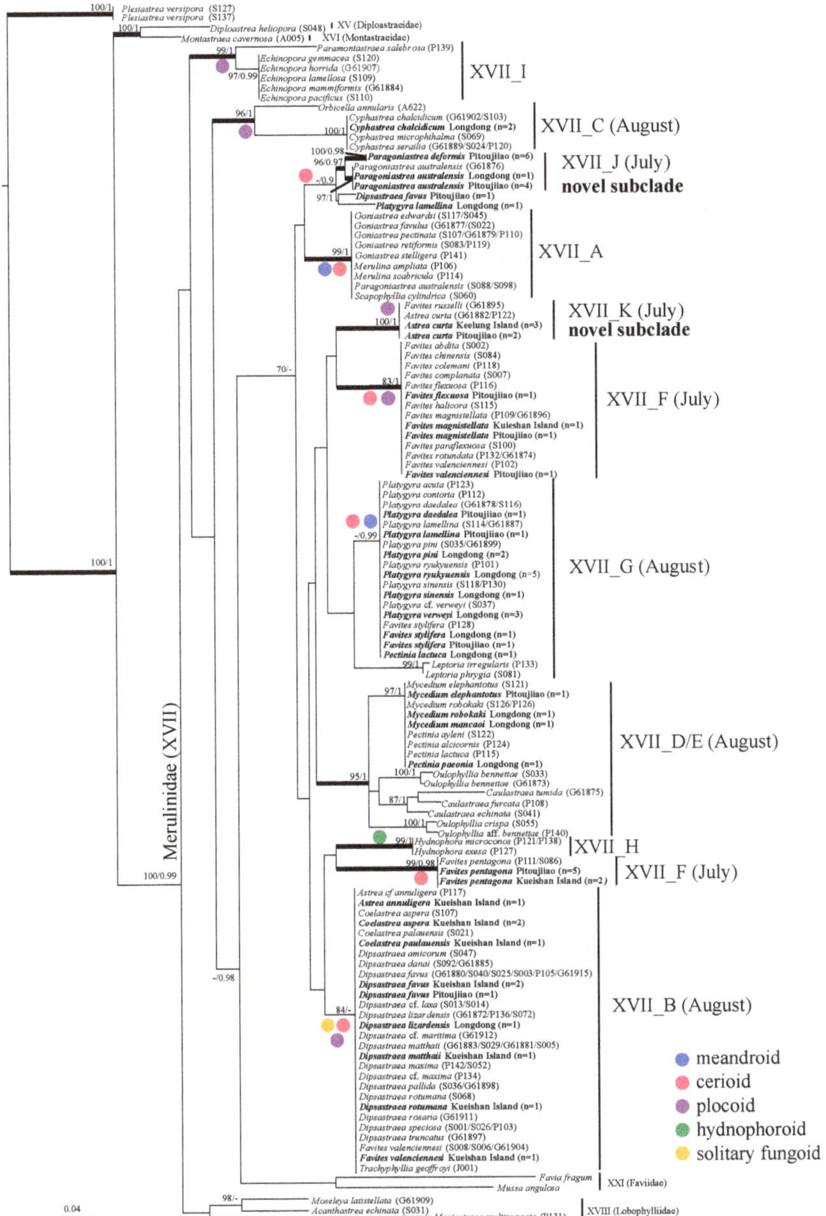

Figure 2. Phylogenetic tree of merulinid corals and their allies based on the combined gene dataset inferred with the maximum likelihood method using the GTR+G model. Molecular subclades within Merulinidae (XVII) are defined as being A to I following Huang et al. [10]. The other two novel clades (J and K) are defined in this study. Branch lengths are proportional to inferred nucleotide substitutions. Numbers at the nodes represent bootstrap values (only ≥70 shown) from the maximum likelihood method and posterior probability (only ≥0.9 shown) from the Bayesian inference. Bold branches on the tree indicate statistically robust nodes. The spawning month of specimens in Chen et al. [32] are in brackets.

Favites russilli and *Astrea curta* formed a distinct cluster (BP:100, PP:1), defined as a new subclade K, separated from *Astrea annuligera*, *Paragoniastrea australensis*, and *P. deformis* were monophyletic (BP:96, PP:0.97), so we defined the genus as a new subclade J. *Paragoniastrea australensis*, not monophyletic, was placed in subclade A and new subclade J. *F. valenciennesi*, not monophyletic, was placed in subclades B and F.

3.4. The Phylogenetic Tree

The spawning specimens we examined were all nested within Merulinidae (taxa in bold font in Figure 2). These specimens were placed in five subclades B, C, E, F, and G but not in the subclades A, D, H, or I. The following were nested in subclade B: one specimen each of *Astrea annuligera*, *Coelastrea palauensis*, *Dipsastraea rotumana*, *Dipsastraea mathaii*, and *Favites valenciennesi* and two specimens of *Coelastrea aspera* collected from Kueishan Island; one specimen of *Dipsastraea favus* collected from Pitoujiiao and two from Kueishan Island; one specimen of *Dipsastraea lizardensis* collected from Longdong. Two *Cyphastrea chalcidicum* specimens, collected from Longdong, were nested in subclade C. One specimen each of *Mycedium mancaoi*, *Mycedium robokaki*, and *Pectinia paeonia*, collected from Longdong, and one *Mycedium elephantotus*, collected from Pitoujiiao, were nested in subclade E. Subclade F consisted of one specimen each of *Favites valenciennesi* and *Favites flexuosa*, collected from Pitoujiiao, and two *Favites magnistellata*, collected from Pitoujiiao and Kueishan Island. All the *Favites pentagona* specimens collected from Kueishan Island, Pitoujiiao, and Longdong were clustered with *F. pentagona* from Singapore and the Philippines. Subclade G consisted of two *Platygyra pini*, five *Platygyra ryukyuensis*, three *Platygyra verweyi*, and one *Platygyra sinensis*, collected from Longdong; one species each of *Platygyra lamellina* and *Platygyra daedalea*, collected from Pitoujiiao; and two *Favites stylifera* from Pitoujiiao and Longdong. Six *Paragoniastrea deformis*, collected from Pitoujiiao, four *Paragoniastrea australensis* specimens from Pitoujiiao, and one from Longdong were clustered into subclade J. Subclade K was a monophyletic clade, consisting of three specimens of *Astrea curta*, collected from Keelung Island, and two from Pitoujiiao, which were clustered with those from the Great Barrier Reef and the Philippines (BP:98, PP:1).

4. Discussion

This is the first study to establish a molecular database for spawning corals, an important contribution to our understanding of genetic diversity in coral communities. We sequenced 1 species from Keelung Island, 9 from Kueishan Island, 13 from Longdong, and 12 from Pitoujiiao. In total, we sequenced 9 genera and 28 species, and most subclades were consistent with those of previous studies.

4.1. Phylogentic Relationship of Merulinid Subclades

The Merulinidae are defined as monophyletic in this study, confirming previous findings [10,16]. Of four Atlantic species, *Favia fragum*, *Mussa angulosa*, *Orbicella annularis*, and *Montastraea cavernosa*, only *O. annularis* is nested within the subclade C and a sister to *Cyphastrea* spp. The genus *Paramontastraea* Huang et al. 2014 [10] examined in this study was also a sister to *Echinopora* Lamarck, 1816, and nested within subclade I.

Increasing the sequence lengths, taxon sampling, and sampling locations may improve the phylogenetic relationship among taxa. Adding new sequences of merulinid corals from Taiwan generated longer aligned sequences with which to examine the phylogenetic relationships among the subclades of Merulinidae; as a result, most subclades changed their phylogenetic positions (Figure S1). For example, the tree topology reconstructed in Huang et al. [10] showed that *Hydnophora* (subclade H) is closer to *Favites* (subclade F). However, our reconstructed phylogenetic tree showed that the *Hydnophora* lineages only closer to *Favites pentagona* and the rest of *Favites* spp. are close to novel subclade K, which comprises *Astrea curta* and *Favites russelli*. In addition, subclade B shifted its position from D/E clades to subclade H and *Favites pentagona*.

As mentioned in Huang et al. [10,15,16], *Favites pentagona* and *Paragoniastrea autralensis* displayed polyphyletic patterns that require further investigation [10]. *Paragoniastrea australensis* was far from subclade A (*Goniastrea* spp.) and was clustered together with *Astrea curta*, *Astrea annuligera*, *Astrea devatieri*, and *Favites russelli* as a novel clade. *Favites pentagona* renders *Favites* polyphyletic in the molecular phylogeny and sister to the *Favites* spp. (subclade F) and subclade D/E [10]. Therefore, Huang et al. [10,16] suggested that these two species require further study with increasing sample collection from other locations. According to our new analysis, the molecular phylogenetic tree implied that *P. australensis* displayed a polyphyletic pattern, which is consistent with Huang et al. [10,16]. However, they were close to subclade A (*Goniastrea* spp.) and formed a novel subclade J (*Paragoniastrea* spp.). *Favites pentagona* formed a monophyletic pattern, which is different from Huang et al. [10,16]. They are separate from the major *Favites* spp. (subclade F) and are close to subclade H (*Hydnophora*).

4.2. Application of Molecular Phylogentic Approaches

Merulinidae corals with plocoid and ceroid forms are difficult to accurately identify to the genus level in the field because of their macro-morphological homoplasies [24,33]. In our phylogenetic analysis, all of the corals in the plocoid form were placed into subclades B, C, F, I, and/or K, including 18 *Dipsastraea* spp. (subclade B), 2 *Favites* spp. (B and F), 5 *Cyphastrea* spp. (C), and 2 *Astrea* spp. (B and K). Merulinidae corals in the ceroid form included 6 *Goniastrea* spp. (subclade A), 14 *Favites* spp. (F), 2 *Coelastrea* spp. (B), and 9 *Platygyra* spp. (G). Most of the samples we examined were either *Coelastrea* spp., *Favites pentagona*, *Platygyra* spp., *Paragoniastrea australensis*, or *Paragoniastrea deformis*. The genetic divergence between these four groups (subclades B, F, G, and J) may be driven by the differences in their sexually reproductive timing. *Favites pentagona* and *P. deformis* spawn in July, while *Platygyra* spp. and *P. australensis* spawn in August [32].

4.3. Sexual Reproduction in Merulinidae

Scleractinian corals have a complex sexual reproduction system, with the same species displaying different sexual reproduction patterns based on their geographic distribution [45]. The systematic pattern in the reproductive biology of Merulinidae, a hermaphroditic spawner, is highly conserved [45–47]. In this study, *Coelastrea aspera* collected from northern Taiwan was identified as a hermaphroditic spawner and placed in subclade B with the Singapore *Coelastrea aspera* [32,48]. *Coelastrea aspera*, from the Great Barrier Reef, is a spawner [49–52], whereas the nonspecific populations distributed in Palau are brooders [53,54]. The hermaphroditic *Coelastrea aspera*, from Okinawa, performs as a spawner [55] and a brooder [56–58]. A similar example, *Pocillopora damicornis*, is a brooder in most locations, but a spawner in western Australia [59–62].

5. Conclusions

This study integrates reproduction information, morphological characteristics, and molecular phylogenetic analysis to increase our understanding of the genetic diversity of Merulinidae. Ten major subclades (A, B, C, D/E, F, G, H, I, J, and K) were reconstructed. Our study identified *Paragoniastrea deformis* and *Paragoniastrea australensis* in Taiwan for the first time. Together, the two species form the new subclade J. *Astrea curta* were separated from another congeneric species, *Astrea annuligera* (XVII-B), clustered with *Favites russelli* into the new subclade K. Finally, we contributed information on the species diversity of coral communities in Taiwan and fill gaps involving merulinid corals between Japan and the Philippines in the Western Pacific.

Supplementary Materials: The following supporting information can be downloaded at: https://www.mdpi.com/article/10.3390/d14020144/s1, Table S1: Species and DNA sequences examined in this study. The species name in Huang et al. 2011 was updated according to World Register of Marine Species (WoRMs). GenBank accession numbers are displayed for each molecular marker (28S rDNA, histone H3, ITS rDNA and mt COI). N.D.: sequences were failed from PCR. Accession dates of 28S rDNA sequences: 21-FEB-2011 [16], 26-OCT-2021 (this study); histone H3: 25-JUL-2016 [16], 23-NOV-2021 (this study); ITS rDNA: 21-FEB-2011 [16], 04-DEC-2021 (this study). Figure S1. Phylogeny of Merulinidae reconstructed from Huang et al. [10] and this study.

Author Contributions: C.-J.C. and C.-F.C. conceived and designed the experiments. C.-J.C. and Y.-Y.H. the collected samples, conducted the experiments, and analyzed the data. C.-J.C. and C.-F.C. wrote the manuscript. All authors have read and agreed to the published version of the manuscript.

Funding: This study was supported by MOST 104-2923-B-019-MY4, Taiwan.

Institutional Review Board Statement: All procedures and investigations were approved by the College of Life Science of the National Taiwan Ocean University Institutional Animal Care and Use Committee (Affidavit of Approval of Animal Use Protocol: No. 104009) and performed in accordance with standard guiding principles.

Informed Consent Statement: Not applicable.

Data Availability Statement: Data are contained within this article. Raw data are available on request from the corresponding authors.

Acknowledgments: We are grateful to the diving instructors, Chen-Ta Wu and Yi-Hung Li for assisting with night dives during the spawning periods. We thank Pei-Chung Lo from the Marine Biodiversity and Phylogenomics Laboratory at National Taiwan University for the stimulating discussions and insightful comments. The sampling permit was issued by the Fishes and Fishing Port Affairs Management Office, New Taipei City Government. This work was supported by research grants from the Ministry of Science and Technology, Taiwan (MOST) to C.F. Chang. Thanks also to Noah Last for editing the manuscript's English.

Conflicts of Interest: The authors declare no conflict of interest.

References

1. Veron, J.E.N.; Pichon, M. *Scleractinia of Eastern Australia*; Australian Govt. Pub. Service: Townsville, Australia, 1976; Volume 1.
2. Veron, J.E.N. *Corals of the World*; Australian Institute of Marine Science: Townsville, Australia, 2000; Volume 3.
3. Todd, P.A. Morphological plasticity in scleractinian corals. *Biol. Rev. Camb. Philos. Soc.* **2008**, *83*, 315–337. [CrossRef] [PubMed]
4. Wells, J.W. Scleractinia. In *Treatise on Invertebrate Paleontology, Coelenterata*; Teichert, C., Ed.; Geological Society of America and University of Kansas Press: Lawrence, KS, USA, 1956; pp. 328–444.
5. Veron, J.E. *Corals in Space and Time: The Biogeography & Evolution of the Scleractinia*; Cornell University Press: Ithaca, NY, USA, 1995.
6. Romano, S.L.; Palumbi, S.R. Evolution of scleractinian corals inferred from molecular systematics. *Science* **1996**, *271*, 640–642. [CrossRef]
7. Romano, S.L.; Palumbi, S.R. Molecular evolution of a portion of the mitochondrial 16S ribosomal gene region in scleractinian corals. *J. Mol. Evol.* **1997**, *45*, 397–411. [CrossRef] [PubMed]
8. Kitahara, M.V.; Cairns, S.D.; Stolarski, J.; Blair, D.; Miller, D.J. A comprehensive phylogenetic analysis of the scleractinia (Cnidaria, Anthozoa) based on mitochondrial CO1 sequence data. *PLoS ONE* **2010**, *5*, 1490. [CrossRef]
9. Fukami, H.; Chen, C.A.; Budd, A.F.; Collins, A.; Wallace, C.; Chuang, Y.Y.; Chen, C.; Dai, C.F.; Iwao, K.; Sheppard, C.; et al. Mitochondrial and nuclear genes suggest that stony corals are monophyletic but most families of stony corals are not (Order Scleractinia, Class Anthozoa, Phylum Cnidaria). *PLoS ONE* **2008**, *3*, e3222. [CrossRef]
10. Huang, D.; Benzoni, F.; Fukami, H.; Knowlton, N.; Smith, N.D.; Budd, A.F. Taxonomic classification of the reef coral families Merulinidae, Montastraeidae, and Diploastraeidae (Cnidaria: Anthozoa: Scleractinia). *Zool. J. Linnean Soc.* **2014**, *171*, 277–355. [CrossRef]
11. Kuzon, K.S.; Lin, M.F.; Lagman, M.C.A.A.; Licuanan, W.R.Y.; Chen, C.A. Resurrecting a subgenus to genus: Molecular phylogeny of *Euphyllia* and *Fimbriaphyllia* (order Scleractinia; family Euphyllidae; clade V). *PeerJ* **2017**, *5*, e4074.
12. Benzoni, F.; Arrigoni, R.; Waheed, Z.; Stefani, F.; Hoeksema, B.W. Phylogenetic relationships and revision of the genus Blastomussa (Cnidaria: Anthozoa: Scleractinia) with description of a new species. *Baffles Bull. Zool.* **2014**, *62*, 358–378.
13. Budd, A.F. Systematics and evolution of scleractinian corals. In *Encyclopedia of Life Synthesis Meeting Report*; National Museum of Natural History: Washington, DC, USA, 2009.
14. Budd, A.F.; Stolarski, J. Corallite wall and septal microstructure in scleractinian reef corals: Comparison of molecular clades within the family Faviidae. *J. Morphol.* **2011**, *272*, 66–88. [CrossRef]
15. Huang, D.; Benzoni, F.; Arrigoni, R.; Baird, A.H.; Berumen, M.L.; Bouwmeester, J.; Chou, L.M.; Fukami, H.; Licuanan, W.Y.; Lovell, E.R.; et al. Towards a phylogenetic classification of reef corals: The Indo-Pacific genera *Merulina*, *Goniastrea* and *Scapophyllia* (Scleractinia, Merulinidae). *Zool. Scr.* **2014**, *43*, 531–548. [CrossRef]
16. Huang, D.; Licuanan, W.Y.; Baird, A.H.; Fukami, H. Cleaning up the 'Bigmessidae': Molecular phylogeny of scleractinian corals from Faviidae, Merulinidae, Pectiniidae and Trachyphylliidae. *BMC Evol. Biol.* **2011**, *11*, 37. [CrossRef]

17. Huang, D.W.; Meier, R.; Tood, P.A.; Chou, L.M. More evidence for pervasive paraphyly in scleractinian corals: Systematic study of Southeast Asian Faviidae (Cnidaria; Scleractinia) based on molecular and morphological data. *Mol. Phylogenet. Evol.* **2009**, *50*, 102–116. [CrossRef] [PubMed]

18. Fukami, H.; Budd, A.F.; Paulay, G.; Sollé-Cava, A.; Chen, C.A.; Iwao, K.; Knowlton, N. Conventional taxonomy obecures deep divergence between Pacific and Atlantic corals. *Nature* **2004**, *427*, 832–834. [CrossRef] [PubMed]

19. Schwartz, S.A.; Budd, A.F.; Carlon, D.B. Molecules and fossils reveal punctuated diversification in Caribbean "faviid" corals. *BMC Evol. Biol.* **2012**, *12*, 123. [CrossRef] [PubMed]

20. World List of Scleractinia. Available online: http://www.marinespecies.org/scleractinia (accessed on 2 August 2021).

21. Khalih, H.M.; Fathy, M.S.; Sawy, S.M.A. Quaternary corals (Scleractinia: Merulinidae) from the Egyptian and Saudi Arabian Red Sea Coast. *Geol. J.* **2021**, *56*, 4150–4188. [CrossRef]

22. Veron, J.E.N.; De Vantier, L.M.; Turak, E.; Green, A.L.; Kininmonth, S.; Stafford-Smith, M.; Peterson, N. The Coral Triangle. In *Coral Reefs: An Ecosystem in Transition*; Dubinsky, Z., Stambler, N., Eds.; Springer: Dordrecht, The Netherlands, 2011; pp. 47–55.

23. Dai, C.F.; Soong, K.; Chen, C.A.; Fan, T.Y.; Hsieh, H.J.; Jeng, M.S.; Chen, C.H.; Horng, S. *Status of Coral Reefs of Taiwan*; Institute of Oceanography, National Taiwan University: Taipei, Taiwan, 2009.

24. Ribas-Deulofeu, L.; Denis, V.; Palmass, S.; Kuo, C.Y.; Hsieh, H.; Chen, C. Structure of benthic communities along the Taiwan latitudinal gradient. *PLoS ONE* **2016**, *11*, e0160601. [CrossRef] [PubMed]

25. Dai, C.F.; Cheng, Y.L. *Corals of Taiwan*; Institute of Oceanography: Taipei, Taiwan, 2020; Volume 1.

26. Chen, C.A.; Wallace, C.C.; Yu, J.K.; Wei, N.V. Strategies for amplification by polymerase chain reaction of the complete sequence of the gene encoding nuclear large subunit ribosomal RNA in corals. *Mar. Biotechnol.* **2000**, *2*, 558–570. [CrossRef] [PubMed]

27. Lin, M.F.; Kitahara, M.V.; Tachikawa, H.; Keshavmurthy, S.; Chen, C.A. A new shallow-water species, *Polycyathus chaishanensis* sp. nov. (Scleractinia: Caryophylliidae), from Chaishan, Kaohsiung, Taiwan. *Zool. Stud.* **2012**, *51*, 213–221.

28. De Palmas, S.; Soto, D.; Denis, V.; Ho, M.J.; Chen, C.A. Molecular assessment of *Pocillopora verrucosa* (Scleractinia; Pocilloporidae) distribution along a depth gradient in Ludao, Taiwan. *PeerJ* **2018**, *6*, e5797. [CrossRef]

29. Soto, D.; De Palmas, S.; Ho, M.J.; Denis, V.; Chen, C.A. Spatial variation in the morphological traits of *Pocillopora verrucosa* along a depth gradient in Taiwan. *PLoS ONE* **2018**, *13*, e0202586. [CrossRef]

30. Hoeksema, B.W.; Arrigoni, R. DNA barcoding of a stowaway reef coral in the international aquarium trade results in a new distribution record. *Mar. Biodivers.* **2020**, *50*, 41. [CrossRef]

31. Shikina, S.; Chang, C.F. Sexual Reproduction in Stony Corals and Insight into the Evolution of Oogenesis in Cnidaria. In *The Cnidaria, Past, Present and Future*; Goffredo, S., Dubinsky, Z., Eds.; Springer: Cham, Switzerland, 2016; pp. 249–268.

32. Chen, C.J.; Chen, W.J.; Chang, C.F. Multispecies spawning of scleractinian corals in non-reefal coral communities of northern Taiwan in the northwestern Pacific Ocean. *Bull. Mar. Sci.* **2021**, *97*, 351–371. [CrossRef]

33. Lin, Y.V.; Denis, V. Acknowledging differences: Number, characteristics, and distribution of marine benthic communities along Taiwan coast. *Ecosphere* **2019**, *10*, e02803. [CrossRef]

34. Madden, T. *The NCBI Handbook*; National Center for Biotechnology: Bethesda, MD, USA, 2002.

35. Katoh, K.; Standley, D.M. MAFFT multiple sequence alignment software version 7: Improvements in performance and usability. *Mol. Biol. Evol.* **2013**, *30*, 772–780. [CrossRef]

36. Rambaut, A. Se-Al Sequence Alignment Editor, Version 2.0a11. Available online: http://tree.bio.ed.ac.uk/software/seal/ (accessed on 8 August 2002).

37. Sworfford, D.L. *PAUP: Phylogenetic Analysis Using Parsimony, Version 3.1*; Illinois Natural History Survey: Champaign, IL, USA, 1991.

38. Miller, M.A.; Pfeiffer, W.; Schwartz, T. Creating the CIPRES Science Gateway for Inference of Large Phylogenetic Trees. In *2010 Gateway Computing Environments Workshop (GCE)*; IEEE: Piscataway, NJ, USA, 2010.

39. Edler, D.; Klein, J.; Antonelli, A.; Silvestro, D. raxmlGUI 2.0: A graphical interface and toolkit for phylogenetic analyses using RAxML. *Methods Ecol. Evol.* **2021**, *12*, 373–377. [CrossRef]

40. Felsenstein, J. Evolutionary trees from DNA sequences: A maximum likelihood approach. *J. Mol. Evol.* **1981**, *17*, 368–376. [CrossRef]

41. Ronquist, F.; Teslenko, M.; van der Mark, P.; Ayres, D.L.; Darling, A.; Höhana, S.; Larget, B.; Liu, L.; Suchard, M.A.; Huelsenbeck, J.P. Efficient Bayesian phylogenetic inference and model choice across a large model space. *Syst. Biol.* **2012**, *61*, 539–542. [CrossRef]

42. Lanfear, R.; Calcott, B.; Ho, S.Y.; Guindon, S. Partitionfinder: Combined selection of partitioning schemes and substitution models for phylogenetic analyses. *Mol. Biol. Evol.* **2012**, *29*, 1695–1701. [CrossRef]

43. Rambaut, A.; Suchard, M.A.; Xie, D.; Drummond, A.J. Tracer v1.6. Available online: http://beast.bio.ed.ac.uk/Tracer (accessed on 30 November 2009).

44. Rambaut, A. FigTree v1.4.2. Available online: http://tree.bio.ed.ac.uk/software/figtree/ (accessed on 10 March 2019).

45. Harrison, P.L. Sexual Reproduction of Scleractinian Corals. In *Coral Reefs: An Ecosystem in Transition*; Dubinsky, Z., Stambler, N., Eds.; Springer: Dordrecht, The Netherlands, 2011; pp. 59–85.

46. Kerr, A.M.; Baird, A.H.; Hughes, T.P. Correlated evolution of sex and reproductive mode in corals (Anthozoa: Scleractinia). *Proc. R. Soc. B* **2010**, *278*, 75–81. [CrossRef]

47. Baird, A.H.; Guest, J.R.; Willis, B.L. Systematic and biogeographical patterns in the reproductive biology of scleractinian coral. *Annu. Rev. Ecol. Evol.* **2009**, *40*, 551–571. [CrossRef]

48. Dai, C.F.; Soong, K.; Fan, T.Y. Sexual reproduction of corals in northern and southern Taiwan. In Proceedings of the International Coral Reef Symposium (ICRS), Guam, Micronesia, 22–27 June 1992; Volume 1, pp. 448–455.

49. Babcock, R. Reproduction and distribution of two species of *Goniastrea* (Scleractinia) from the Great Barrier Reef Province. *Coral Reefs* **1984**, *2*, 187–195.

50. Babcock, R.C.; Bull, G.D.; Harrison, P.L.; Heyward, A.J.; Oliver, J.K.; Wallace, C.C.; Willis, B.L. Synchronous spawnings of 105 scleractinian coral species on the Great Barrier Reef. *Mar. Biol.* **1986**, *90*, 379–394. [CrossRef]

51. Harrison, P.L.; Babcock, R.C.; Bull, G.D.; Oliver, J.K.; Wallace, C.C.; Willis, B.L. Mass spawning in tropical reef corals. *Science* **1984**, *223*, 1186–1189. [CrossRef] [PubMed]

52. Willis, B.L.; Babcock, R.C.; Harrison, P.L.; Oliver, J.K.; Wallace, C.C. Patterns in the mass spawning of corals on the Great Barrier Reef from 1981 to 1984. In Proceedings of the Fifth International Coral Reef Congress, Tahiti, French Polynesia, 27 May–1 June 1985; pp. 343–348.

53. Abe, N. Postlarval development of the coral *Fungia actiniformis* var. palawensis Doderlein. *Palao Trop. Biol. Stat. Stud.* **1937**, *1*, 73–93.

54. Motoda, S. Observation of period of extrusion of planula of *Goniastrea aspera* (Verrill.). *Kagaku Nanyo* **1939**, *1*, 5–7.

55. Sakai, K. Gametogenesis, spawning, and planula brooding by the reef coral *Goniastrea aspera* (Scleractinia) in Okinawa, Japan. *Mar. Ecol. Prog. Ser.* **1997**, *151*, 67–72. [CrossRef]

56. Heyward, A.K. Sexual reproduction of corals in Okinawa. *Galaxea* **1987**, *6*, 331–343.

57. Hayashibara, T.; Shimoike, K.; Kimura, T.; Hosaka, S.; Heyward, A.; Harrison, P.L.; Kudo, K.; Omori, M. Patterns of coral spawning at Akajima Island, Okinawa, Japan. *Mar. Ecol. Prog. Ser.* **1993**, *101*, 253–262. [CrossRef]

58. Nozawa, Y.; Harrison, P.L. Temporal settlement patterns of larvae of the broadcast spawning reef coral *Favites chinensis* and the broadcast spawning and brooding reef coral *Goniastrea aspera* from Okinawa, Japan. *Coral Reefs* **2005**, *24*, 274–282. [CrossRef]

59. Castrillón, A.; Muñoz, C.; Zapata, F. Reproductive patterns of the coral *Pocillopora damicornis* at Gorgona Island, Colombian Pacific Ocean. *Mar. Biol. Res.* **2015**, *15*, 1065–1075. [CrossRef]

60. Richmond, R.H.; Jokiel, P.L. Lunar periodicity in larva release in the reef coral *Pocillopora damicornis* at Enewetak and Hawaii. *Bull. Mar. Sci.* **1984**, *34*, 280–287.

61. Stoddart, J.A.; Black, R. Cycles of gametogenesis and planulation in the coral *Pocillopora damicornis*. *Mar. Ecol. Prog. Ser.* **1985**, *23*, 153–164. [CrossRef]

62. Ward, S. Evidence for broadcast spawning as well as brooding in the scleractinian coral *Pocillopora damicornis*. *Mar. Biol.* **1992**, *112*, 641–646. [CrossRef]

Diversity of Coral Reef Fishes in the Western Indian Ocean: Implications for Conservation

Melita Samoilys [1,2,*], Lorenzo Alvarez-Filip [3], Robert Myers [4] and Pascale Chabanet [5]

[1] CORDIO East Africa, P.O. Box 10135, Mombasa 80101, Kenya
[2] Department of Zoology, University of Oxford, Oxford OX1 3PS, UK
[3] Biodiversity and Reef Conservation Laboratory, Unidad Académica de Sistemas Arrecifales, Instituto de Ciencias del Mar y Limnología, Universidad Nacional Autónoma de México, Puerto Morelos 4510, Mexico; lorenzoaf@gmail.com
[4] Seaclicks/Coral Graphics, 33411 Wellington, FL, USA; robmyers1423@gmail.com
[5] UMR ENTROPIE (IRD, UR, CNRS, IFREMER, UNC), CS 41096, 2 Rue Joseph Wetzell, 97495 La Reunion, France; pascale.chabanet@ird.fr
* Correspondence: msamoilys@cordioea.net; Tel.: +254-721498713

Abstract: Communities of coral reef fishes are changing due to global warming and overfishing. To understand these changes and inform conservation, knowledge of species diversity and distributions is needed. The western Indian Ocean (WIO) contains the second highest coral reef biodiversity hotspot globally, yet a detailed analysis of the diversity of coral reef fishes is lacking. This study developed a timed visual census method and recorded 356 species from 19 families across four countries in the WIO to examine patterns in species diversity. Species richness and composition differed most between the island countries of Madagascar and Comoros and both these locations differed from locations in Tanzania and Mozambique which were similar. These three regional groupings helped define WIO ecoregions for conservation planning. The highest species richness was found in Tanzania and Mozambique, and the lowest and most different species composition was found in Comoros. Biogeography explains these differences with naturally lower species diversity expected from the small, oceanic, and isolated islands of Comoros. Present day ocean currents maintain these diversity patterns and help explain the species composition in northeast Madagascar. Species distributions were driven by 46 of the 356 species; these provide guidance on important species for ongoing monitoring. The results provide a benchmark for testing future changes in reef fish species richness.

Keywords: coral reef fishes; species diversity; Indian Ocean; conservation; climate change

Citation: Samoilys, M.; Alvarez-Filip, L.; Myers, R.; Chabanet, P. Diversity of Coral Reef Fishes in the Western Indian Ocean: Implications for Conservation. *Diversity* **2022**, *14*, 102. https://doi.org/10.3390/d14020102

Academic Editor: Simone Montano

Received: 13 December 2021
Accepted: 26 January 2022
Published: 31 January 2022

Publisher's Note: MDPI stays neutral with regard to jurisdictional claims in published maps and institutional affiliations.

1. Introduction

Species are the fundamental units of ecosystems and thus species inventories and their distributions provide a foundation for understanding coral reef communities and their conservation [1,2]. Communities of reef-associated fish species reflect their biogeography, and this includes evolutionary history, sea surface temperature, and larval recruitment patterns driven by ocean currents [1,3–5]. But these reef fish communities are changing due to global warming and overfishing, which are rapidly degrading coral reefs globally [6–8], driving declines in abundance and local extirpations of some species [9,10]. Understanding these changes and informing conservation knowledge about patterns in species diversity is needed.

Marine provinces, first defined over 150 years ago [11,12], along with barriers to species dispersal [13,14], provide a framework for understanding present day biogeographic patterns. The western Indian Ocean (WIO), is considered a distinct province of the Indo–Pacific region [15,16] and comprises 10 countries, all with coral reefs [17,18]. This province contains the second highest biodiversity hotspot in the Indo–Pacific, second to the Coral Triangle in the western Pacific [19–21]. Ten biogeographic subregions within the WIO Province were defined using hermatypic corals [22], with the diversity hotspot centered in the northern Mozambique Channel on the coasts of northern Madagascar, the Comoros Archipelago, northern Mozambique, and southern Tanzania (Figure 1), an area considered likely to host the highest diversity and abundance of other marine fauna [23]. Veron and coauthors [24], based on zooxanthellate coral distributions, confirmed similar coral ecoregions in the WIO but delineated 12 subregions by separating Comoros from Mozambique and Madagascar. These ecoregions provide important conservation planning units [25].

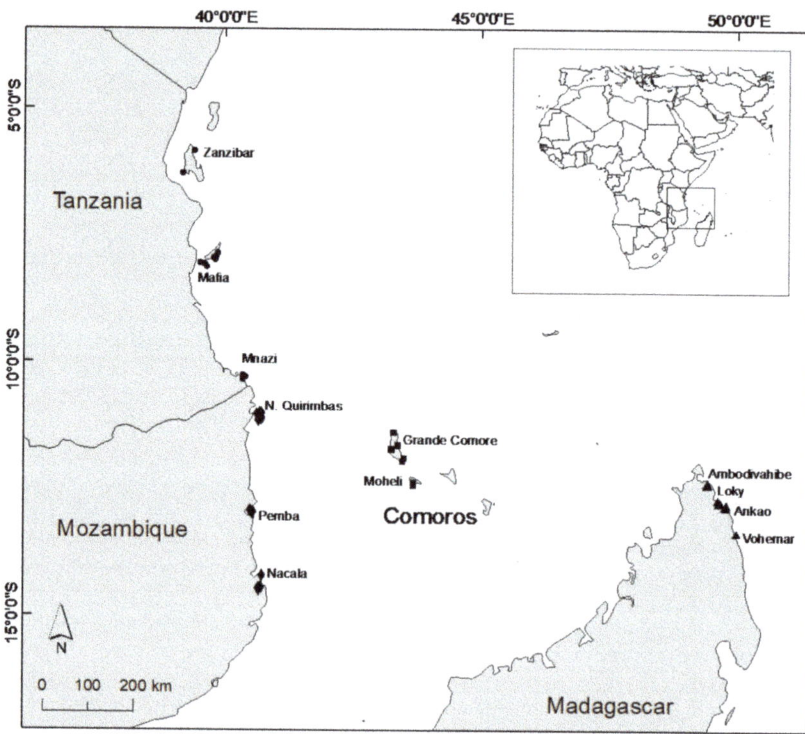

Figure 1. Map of study area showing countries and location of survey sites.

There are over 3000 tropical reef fish species found in the Indian Ocean, of which 74% range widely through the Indo–Pacific, thus giving ~850 species restricted to the Indian Ocean [5]. The WIO Province supports just over 2400 fish species, representing a second peak in fish diversity in the Indo–Pacific after the Coral Triangle [19]. The highest fish species richness is found to the west on the eastern African continental coastline, with ~600 to 960 species. The highest level of endemism in the WIO is found to the east in the Mascarene Islands of Reunion and Mauritius. High endemism is typical of peripheral biogeographic regions [15,26]. Reef fish species' inventories in the WIO remain scattered and largely at a national level ranging from Madagascar [27,28], Comoros [29], Iles Eparses [30,31], to Reunion Island [32]. Therefore, data on species ranges are incomplete, though considerable early work established a sound base of identification sources [33,34]. However, a detailed regional analysis of the diversity of coral reef fishes in the WIO is lacking.

The present study developed a rapid underwater visual census method to compile reef fish species inventories across shallow reefs to examine biogeographic patterns in species assemblages across four countries. The study used the most diverse and/or most numeric families that occur on coral reefs in the Indo–Pacific [35], representing potentially around 460 coral reef species from the WIO [3,27,34]. Families selected represented those reported as indicators of biogeographical patterns and coral reef health, such as Chaetodontidae [36,37]; of fishery importance [38]; of both wide-ranging and restricted range species; and of Tetradontiformes, to expand the taxonomic diversity at the suborder level [3,39,40]. Two highly diverse families notably absent from this list are the Gobiidae and Blennidae, which are known to be excellent biodiversity indicators [41], however, they are too cryptic in their behavior and difficult to identify underwater while surveying a broad suite of species that range up to ~1 m in length. The final 19 families rank highly in importance as indicators on coral reefs (Table 1) and comprises a potential species list that represents ~50% of the putative total number of shallow coral reef species in the WIO [5]. This was therefore considered sufficiently broad and diverse to capture biogeographic patterns in the diversity of fishes within the region.

Table 1. Nineteen families surveyed for coral reef fish diversity analyses based on: (a) most speciose; (b) known indicators of aspects of fish communities; (c) fishery importance; (d) taxonomic diversity. Other rankings of coral reef fish families as indicators of diversity or importance on coral reefs are shown for comparison: Coral Reef Fish Diversity Index, CFDI [A] (Allen and Werner 2002); numerical abundance [C and B] (Choat and Bellwood 1991); consensus list of 10 characteristic coral reef families, B (Bellwood 1996); global comparison of most speciose families [B and W] (Bellwood and Wainright 2002). These families are characteristic of coral reefs though not necessarily restricted to them and are among first 28 most speciose families of reef fish worldwide out of a possible 76 families (Bellwood and Wainright 2002).

Order	Suborder	Families	CFDI [A]	Abundance [C and B]	10 Coral Reef [B]	Most Speciose [B and W]
Perciformes (a)	Labroidei	Labridae (wrasse)	X	X	X	1
	Percoidei	Epinephelidae (groupers)				3
	Labroidei	Pomacentridae (damsel fishes)	X	X	X	2
Perciformes (b)	Percoidei	Chaetodontidae (butterfly fishes)	X	X	X	6
	Labroidei	Scarinae (parrot fishes) [1]	X		X	8
	Acanthuroidei	Acanthuridae (surgeon fishes)	X	X	X	7
	Percoidei	Lutjanidae (snappers)				10
	Percoidei	Pomacanthidae (angel fishes)	X			11
Perciformes (c)	Percoidei	Lethrinidae (emperors)				13
	Percoidei	Haemulidae (grunts)				23
	Percoidei	Mullidae (goat fishes)			X	19
	Acanthuroidei	Siganidae (rabbit fishes)				21
	Percoidei	Nemipteridae (bream)				24
	Percoidei	Carangidae (trevally)			X	N/A
Perciformes (d)	Percoidei	Caesionidae (fusiliers)				28
Tetraodontiformes	Tetraodontiformes	Balistidae (trigger fishes)				16
Tetraodontiformes	Tetraodontiformes	Monacanthidae (file fishes)				14
Tetraodontiformes	Tetraodontiformes	Ostraciidae (box fishes)				25
Tetraodontiformes	Tetraodontiformes	Tetraodontidae (puffer fishes)				18

[1] Scarinae are a subfamily within the Labridae (Bellwood et al. 2019) but for functional purposes are treated separately.

The current study aimed to examine patterns in species richness of reef fishes to contribute to our understanding of the biogeography of the less studied WIO province. It also aimed to assess how reef fish diversity patterns conform to known biodiversity hotspots and to delineate ecoregions in the WIO for coral reef conservation planning and threat assessments.

2. Materials and Methods

2.1. Study Sites

Coral reef fish species were recorded in 2009–2011 at 76 dive stations aligned to 45 sites which ranged from 1–33 m in depth (in the supplementary Table S1), spread across locations in four countries in the WIO: Madagascar, Comoros, Mozambique, and Tanzania (Figure 1) spanning latitude −5.84° (Zanzibar) to −14.47° (Nacala) and longitude 39.17 (Chumbe) to 50.01 (Vohemar). Sites surveyed in Comoros were in Ngazidja (Grand Comore) and Mwali (Moheli), two of the three islands in the Union of Comoros, which are referred to hereafter as Comoros. The fourth southeastern island in the Comoros Archipelago, Mayotte, an overseas department of France, is a larger island with considerably more reef habitat, but was not surveyed here. Each country's dataset is a sample and cannot claim to be representative of the country as a whole. Sites were selected haphazardly and ranged from shallow, protected fringing reefs to deep, exposed forereefs (Table S1). Each location encompassed sites across a large depth range and were therefore broadly comparable, though this negated any analysis for habitat effects (Table S1). Forereefs and deep and shallow terraces were prioritized as these reef types tend to have higher coral cover and rugosity, and hence, higher fish species diversity.

2.2. Survey Method

Coral reef fish diversity was measured by recording presence/absence of species on a SCUBA based underwater visual census (UVC) survey, which involved a timed swim by one observer throughout (MS), recording all species within visibility (mean 14.1 m), supplemented with a few snorkel dives in shallow waters. The method was designed to provide sufficient breadth of species sampling, while remaining practically feasible for relatively rapid dive surveys across a large number of locations. The 19 families were selected based on those that are most speciose, are amenable to UVC (diurnal and not cryptic), have fishery importance, and included four Tetraodontiform families to extend the taxonomic diversity of the dataset (Table 1).

A complete species inventory of 19 families (Table 1) was recorded at each dive. Dive time and species richness were significantly correlated (R^2 = 0.19, slope = 0.27x + 91, $p < 0.01$) though taxonomic diversity was not ($R^2 < 0.01$, slope = $-0.001x + 63$, $p < 0.86$). The species richness curve showed a species plateau at ≥ 75 min dive duration. Consequently, data from the two replicate dive stations at each site were combined to ensure that each reef site was represented by 75–85 min of underwater observations. This conforms to recommended dive times of 60–90 min [27].

Species identifications were checked using photographs, taxonomic references, and photographic guides (see Table S2 for species list and references). Species names were verified from the online Catalog of Fishes [42].

2.3. Data Analyses

To assess overall patterns of species diversity, the species presence/absence data per reef site was used to calculate the total number of species and the average taxonomic distinctness (D+). The average taxonomic distinctness is a measurement of the average taxonomic path length between two randomly chosen species in the assemblage [43] and was increasingly applied because it is considered a good proxy of biodiversity and it is relatively independent of sampling effort [44]. Species richness and taxonomic distinctness were tested for differences between areas within countries and dive time. Species accumulation curves derived from the Michaelis–Menten index using 9999 permutations [43] were utilized to predict maximum species richness for each country. The Michaelis–Menten equation was chosen because it is independent of the rarity of a species, being based solely on presence. Furthermore, on empirical considerations it was also found to be the least biased by the number of samples and the most stable statistic to use for estimated maximum species richness on previous biogeographic studies in the Indian Ocean [22]. Differences between the predicted maximum species richness from the Michaelis–Menten equation and actual number of species observed in each country were compared using Chi-Square.

To detect similarity patterns in the species assemblages on reefs across the region the Bray–Curtis similarity index was calculated among pairs of sites, and the similarity coefficients used to run two ordination techniques to detect similarity patterns between sites [43]: a cluster analysis and a nonmetric multidimensional scaling (MDS). Both methods were applied as they offer complementary information. The significance of the differences between the geographical areas revealed in the cluster dendrogram and MDS plots was tested with an analysis of similarity (ANOSIM) based on randomization of the similarity matrix [43]. To investigate the main species that account for the observed patterns in species richness across the region, a similarity of percentage analysis (SIMPER) was used to detect the representative species of each geographical zone. Since the ANOSIM results showed significant country differences between Comoros and Madagascar, between these island countries and the mainland countries of Tanzania and Mozambique, and a marginal difference between the two mainland countries, the latter were combined as one, "mainland", for the SIMPER analysis.

3. Results

A total of 356 species from the 19 families (Table S2) were recorded from 45 sites across the four countries. Of these, 15 could not be identified to species but were either recorded to genus (10) or to a species it closely resembled from the Pacific Ocean (5). In all cases these uncertain species could be reidentified on subsequent surveys at different sites by the same observer (MS), and therefore, all were used in the analyses of species diversity.

Overall, the predicted number of species did not differ statistically from the observed number (Figure 2; Chi-square = 0.602; df = 3; p = 0.89). However, on a country-by-country basis, there was a marginally significant difference for Comoros (Chi-square = 3.9; df = 1; p = 0.049). This suggests that surveys from all locations in all four countries were adequate in providing representative values of total species richness, but there was some indication that an increase in the number of surveys in Comoros would improve the data. The results show, based on the 19 families, a predicted mean total number of species of 321 for Mozambique and 319 species for Tanzania, both higher than Comoros (267) and Madagascar (294) (Figure 2). Note that these numbers refer to locations surveyed in each country, and not the country as a whole (see Methods). Mean species diversity per site was highest in Mozambique followed closely by Tanzania, and lowest from sites in Comoros (Figure 3a). In contrast, taxonomic distinctness was highest in Comoros (Figure 3b). However, variability within countries was very high, for example, the highest species richness was recorded at Vamizi lagoon (152) and the lowest at Neptunes (82), both in northern Quirimbas, Mozambique (Supplementary Material Figure S1).

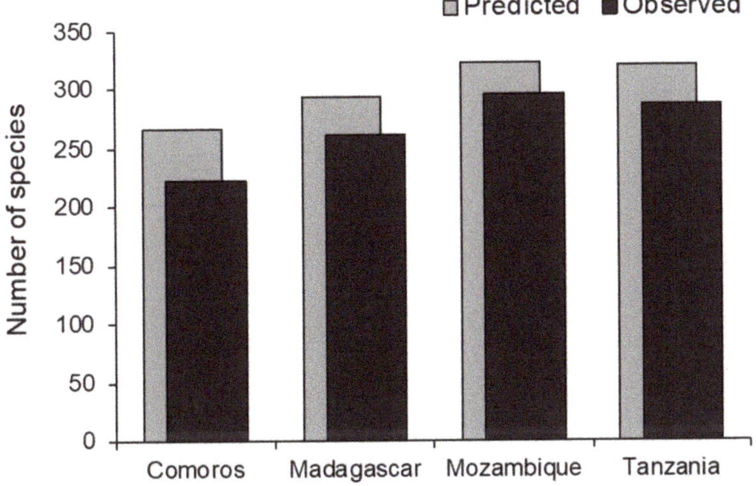

Figure 2. Total number of species observed (black) and predicted maximum species richness (grey) per country based on Michaelis–Menton permutations. Observed and predicted number did not differ significantly (Chi-square = 0.602; p = 0.89). Number of sites: Comoros: 7; Madagascar: 10; Mozambique: 16; Tanzania:12; 76 dives in total, as detailed in Table S1.

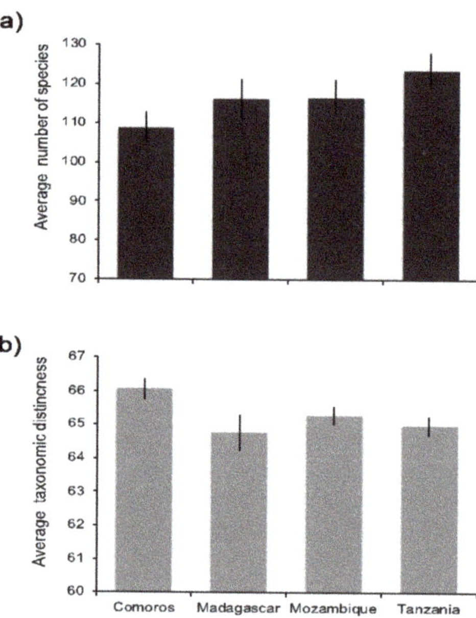

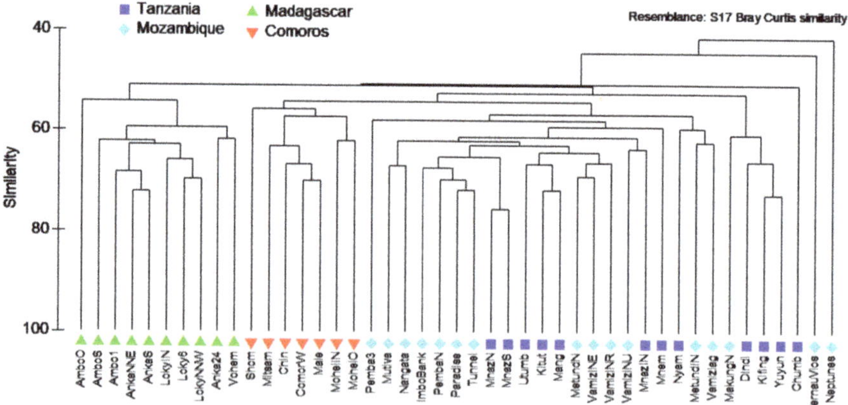

Figure 3. Mean fish species diversity: (**a**) mean total number of species observed per site (>75 min) in each country; (**b**) mean fish species diversity based on taxonomic distinctness (D+). Error bars are standard errors of mean. Sample sizes as in Figure 2.

Similarity in species' presence/absence between sites revealed four groupings at 54% similarity with all the Madagascar sites separate; the Comoros sites clustered separately; most of the sites from the African mainland, Tanzania, and Mozambique, clustered together in a complex set of smaller groups; and three extreme outliers of two sites from Mozambique and one from Tanzania (Figure 4). The same four groupings were confirmed in an MDS ordination though the stress level of 0.2 suggests country differences were not significant (Figure S2). The overall differences in species presence between the four countries were statistically different, with differences between sites in the two island countries, and between sites in island countries and mainland countries all highly significant, whereas differences between Tanzania and Mozambique were only marginally significant (Table 2).

Figure 4. Bray Curtis cluster analysis showing similarity in species presence/absence between pairs of sites in western Indian Ocean (WIO).

Table 2. Results of ANOSIM based on 999 permutations for differences in species richness between (**a**) countries and (**b**) between areas within mainland countries, Tanzania, and Mozambique (based on 999 permutations). Areas within mainland countries that differed significantly are bolded. Pemba is in Cabo Delgado, Mozambique.

(a) Between Countries		
Global R = 0.405, p = 0.001		
Pairwise Tests—Groups	**R Statistic**	p
Madagascar, Comoros	0.864	0.001
Madagascar, Tanzania	0.592	0.001
Madagascar, Mozambique	0.483	0.001
Comoros, Tanzania	0.397	0.001
Comoros, Mozambique	0.354	0.007
Tanzania, Mozambique	0.103	0.048
(b) Between Areas within Mainland Countries		
Global R = 0.21; p = 0.011		
Pairwise Tests—Groups	**R Statistic**	p
Chumbe, Mafia	0.617	0.056
Chumbe, Nacala	0.321	0.200
Chumbe, Pemba	0.857	0.067
Chumbe, Vamizi	0.3	0.156
Chumbe, Mnazi	0.583	0.100
Mafia, Nacala	**0.418**	**0.012**
Mafia, Pemba	**0.489**	**0.009**
Mafia, Vamizi	0.131	0.061
Mafia, Mnazi	0.173	0.225
Nacala, Pemba	−0.01	0.571
Nacala, Vamizi	0.21	0.113
Nacala, Mnazi	0.148	0.229
Pemba, Vamizi	0.104	0.240
Pemba, Mnazi	**0.5**	**0.029**
Vamizi, Mnazi	−0.113	0.636

Fish species assemblages within the sites of the mainland countries, which ranged from −5° (Zanzibar—Mnemba) to −14° (Nacala sites with Fernau Vloso the southernmost), were different but similarity levels were relatively high, and there was no clear latitudinal or geographic pattern, though two outlier sites in Mozambique were apparent (Figures 4 and 5). None of these areas were statistically different from each other in terms of species presence except between two areas in Mozambique (Pemba, Nacala) and two areas in Tanzania (Mafia, Mnazi) (Table 2b). Though not statistically dissimilar (Figure 5), local scale differences in species richness suggested the following groupings (see Table S1 for reef types): (i) exposed forereefs at Mafia Island (Dindini, Yuyuni and Kifinge), similar to Makunga North, which had an exposed reef terrace, and a steeply sloping forereef in northern Mozambique; (ii) inner seas protected forereefs and lagoonal sites at Vamizi and Metundo islands in Mozambique; (iii) the largest group of similar reef sites ranging across the entire east African mainland with well-developed forereefs at ocean-exposed sites; (iv) outliers seen in Zanzibar: Chumbe, which is a narrow and relatively shallow (3–13 m) protected forereef with much sand and low rugosity; Mnemba, which is a relatively deep (to 18 m) exposed forereef; and Nyamlile, a large patch reef off Mafia island. The significantly dissimilar extreme outliers in the whole plot (Figure 5) were two sites in Mozambique: Fernau Vlos in Nacala, a diffuse fringing forereef within the port channel with a few scattered corals on a sandy slope with some seagrass and little hard substrate, and Neptunes, which is an offshore deep terrace between Metundo and Vamizi islands with 90-degree walls dropping to around 500 m. The walls are broken in places with canyons, and the upper terrace ranges from ~7–12 m in depth.

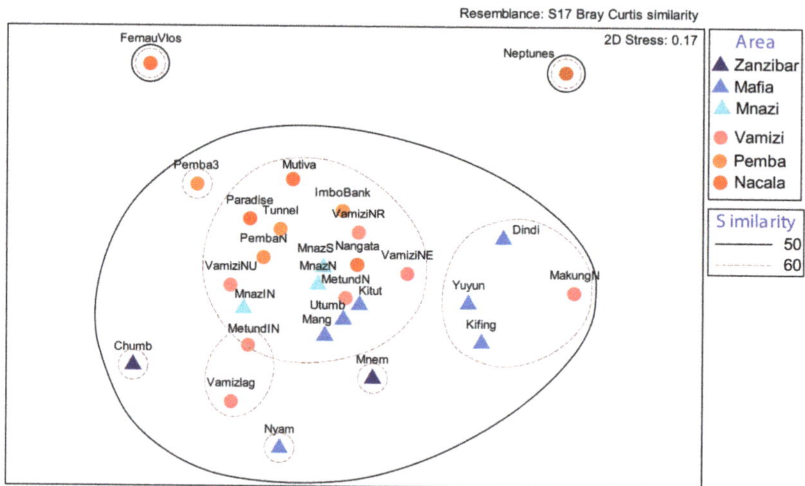

Figure 5. Multidimensional scaling (MDS) plot of mainland sites only, color coded with blues for northern sites (Tanzania) and reds for southern sites (Mozambique). Plot shows all sites are similar at 50%, but within this large group three groups were apparent, though with similarity at 60% these groups were not strongly dissimilar.

Fish species composition and homogeneity were most dissimilar between sites in NE Madagascar and Comoros: of the first 14 high ranking species that contributed to the dissimilarity between all three geographical zone comparisons, 11 contributed most to the differences between Madagascar and Comoros, while the other three high ranking species, *Pygoplites diacanthus*, *Heniochus acuminatus*, and *Cephalopholis argus*, were significant in the other geographic zone comparisons (Table 3). Thus, the SIMPER results of species' average dissimilarity values are ranked according to the top 30 species that most explain the differences between Madagascar and Comoros (Table 3). These cumulatively explain 19.74% of the dissimilarity in the species assemblages between these two geographic zones, and these species contributed between 56% (*Cheilinus trilobatus*) to 85% (*Plectropomus punctatus*) of the dissimilarity in this pairwise comparison (Table S3). The 10 most significant species in each of the three comparisons when combined gives 22 species in total. The average abundance shows that the species that ranked mostly highly had either, very high (0.9–1) relative abundance at sites (they were at most or all sites), or low or zero abundance (they were rare or absent at all sites) (Table 3).

The Madagascar—mainland pairwise comparison yielded many of the same species in explaining differences between assemblages as those seen in the Comoros–Madagascar comparison, but with three notable additions: *Chaetodon falcula*, *Cephalopholis argus*, and *Amphiprion allardi* (the latter two also significant in the Comoros–mainland comparison), as well as an additional eight species specific to Madagascar-mainland (*Acanthurus xanthopterus* to *Calotomus carolinus*, Table 3). In contrast, the Comoros-mainland comparison yielded several different species that did not rank highly in the other two SIMPER comparisons, notably *Heniochus acuminatus*, *Aprion viriscens* and *Ostracion meleagris*. Complete sets of all three pair-wise SIMPER geographic zone comparisons are provided in Table S3.

Table 3. SIMPER results for most significant 47 species, showing first 30 species (*Plectropomus punctatus* to *Cheilinus trilobatus*) that explain 19.74% of dissimilarity in species present at sites according to Madagascar–Comoros dissimilarity cumulative ranking (grey columns), and 17 additional species representing those within top 20 species in other comparisons (12–14% of cumulative mean dissimilarity/SD): (i) Madagascar–mainland comparison, and (ii) Comoros–mainland comparison. Rank AvDiss = Dissimilarity/SD averaged across all three group comparisons is presented as a metric for ranking species that contributed most to differences across all three pair-wise comparisons. Twenty-two species in bold represent top 10 most significant species in all three pair-wise comparisons. Ranges (Catalog of Fishes 2021): IP = Indo–Pacific (includes west coast of Americas); IWP = Indo–West Pacific; IO = Indian Ocean; WIO = western Indian Ocean (Somalia to Mauritius); RS = Red Sea; EA = East Africa (Kenya, Tanzania, Mozambique, Comoros, Madagascar); Mas= Mascarene islands (Reunion, Mauritius, Rodrigues).

Species	Family/Sub-Fam	Range	AVERAGE ABUNDANCE			Diss/SD			Cum.Av. Diss.Contrib %			Rank AvDiss/SD
			COMORC	MAINLAND	MADAGA	Comoros-Mainland	Madagacar-Mainland	Madagascar-Comoros	Comoros-Mainland	Madagacar-Mainland	Madagascar-Comoros	
Plectropomus_punctatus	Epinephelidae	WIO	**0**	**0.32**	**1**	**0.68**	**1.43**	**11.97**	76.92	6.78	0.85	1
Plectorhinchus_gaterinus	Haemulidae	WIO	**0**	**0.57**	**0.9**	**1.14**	**0.88**	**2.89**	13.62	59.79	1.63	5
Hipposcarus_harid	Labridae	RS-IO [1]	**0**	**0.54**	**0.9**	**1.06**	**0.93**	**2.88**	27.36	49.69	2.4	6
Epibulus_insidiator	Labridae	RS-IWP	**0**	**0.79**	**0.9**	**1.87**	**0.61**	**2.88**	0.73	82.37	3.16	3
Chaetodon_vagabundus	Chaetodontidae	IWP	**0**	**0.07**	**0.9**	**0.28**	**2.24**	**2.88**	98.19	2.23	3.93	2
Anampses_twistii	Labridae	RS-IWP	**0.86**	**0.86**	**0**	**0.56**	**2.37**	**2.38**	84.35	1.49	4.66	4
Neoglyphidodon_melas	Pomacentridae	RS-IWP	**0.14**	**0.75**	**1**	**1.43**	**0.57**	**2.37**	2.76	85.29	5.4	8
Scarus_ghobban	Scarinae	RS-IWP	**0.14**	**0.61**	**1**	**1.15**	**0.8**	**2.37**	12.01	66.81	6.12	9
Cheilinus_fasciatus	Labridae	RS-IWP	**0**	**0.61**	**0.8**	**1.23**	**0.87**	**1.95**	7.58	62.1	6.8	11
Acanthurus_dussumieri	Acanthuridae	IWP	**0.14**	**0.21**	**0.9**	**0.64**	**1.6**	**1.87**	80.63	5.57	7.48	10
Chaetodon_melannotus	Chaetodontidae	RS-IWP	**0.14**	**0.75**	**0.9**	**1.43**	**0.65**	**1.87**	3.4	80.32	8.15	13
Pomacanthus_chrysurus	Pomacanthidae	WIO	0.14	0.39	0.9	0.85	1.17	1.87	63.12	12.81	8.82	15
Pomacentrus_baenschi	Pomacentridae	EA	0	0.36	0.8	0.74	1.17	1.96	74.43	14.88	9.49	16
Mulloidichthys_flavolineatus	Mullidae	IWP	0.86	0.36	0.1	1.21	0.78	1.88	8.7	68.86	10.16	17
Parupeneus_trifasciatus	Mullidae	IWP	0.86	0.46	0.1	1.04	0.93	1.87	24.89	54.24	10.83	18
Lutjanus_fulviflamma	Lutjanidae	RS-IWP	0.14	0.64	0.8	1.22	0.83	1.55	9.82	63.6	11.45	23
Pygoplites_diacanthus	Pomacanthidae	RS-IWP	**0.71**	**0.86**	**0**	**0.72**	**2.36**	**1.55**	75.08	0.75	12.06	7
Chlorurus_atrilunula	Scarinae	WIO	0.14	0.43	0.8	0.89	1.07	1.55	57.87	20.73	12.67	25
Thalassoma_genivittatum	Labridae	(WIO-Mas)	**0.14**	**0.07**	**0.8**	**0.49**	**1.72**	**1.55**	89.85	4.92	13.29	21
Chromis_ternatensis	Pomacentridae	IWP	1	0.82	0.3	0.46	1.28	1.5	89.51	9.6	13.9	41
Cheilinus_oxycephalus	Labridae	IWP	**0.14**	**0.75**	**0.8**	**1.43**	**0.73**	**1.55**	4.02	72.46	14.5	22
Abudefduf_vaigiensis	Pomacentridae	RS-IWP	0.71	0.46	0	1.02	0.92	1.55	27.85	59.01	15.11	26
Pomacentrus_trichrourus	Pomacentridae	RS-WIO	0.14	0.54	0.8	1.04	0.95	1.56	26.37	50.95	15.72	24
Pomacentrus_caeruleopunctatus	Pomacentridae	WIO-Mas	**0.14**	**0.04**	**0.8**	**0.45**	**1.83**	**1.55**	92.02	4.27	16.32	20
Lutjanus_ehrenbergi	Lutjanidae	RS-IWP	0	0.11	0.7	0.34	1.36	1.5	96.48	7.35	16.9	47
Pervagor_janthinosoma	Labridae	IWP	0.71	0.21	0.1	1.27	0.6	1.41	7.01	82.85	17.47	32
Coris_formosa	Labridae	RS-WIO [2]	**0.71**	**0.11**	**0.1**	**1.4**	**0.47**	**1.4**	4.64	90.14	18.04	38
Variola_louti	Epinephelidae	RS-IWP	0.71	0.54	0.1	0.96	1.04	1.4	42.1	30.47	18.61	27
Amblyglyphidodon_indicus	Pomacentridae	RS-IO	0.29	0.71	0.9	1.19	0.69	1.41	11.47	77.59	19.17	31
Cheilinus_trilobatus	Labridae	IWP	0.29	0.75	0.9	1.23	0.65	1.4	9.26	79.26	19.74	36
Additional Significant Species (top 20) in Madagascar-Mainland Comparisons												
Cephalopholis_argus	Epinephelidae	RS-IWP	**0.29**	**0.86**	**0.1**	**1.35**	**1.87**	**0.69**	5.25	3.6	81.16	14
Chaetodon_falcula	Chaeotodontidae	IO	**0.43**	**0.86**	**0.1**	**1.09**	**1.87**	**0.88**	16.25	2.91	67.67	19
Amphiprion_allardi	Pomacentridae	WIO	**0**	**0.71**	**0**	**1.55**	**1.55**	**Undefined**	2.12	6.19	100	114
Acanthurus_xanthopterus	Acanthuridae	IWP	0.14	0.11	0.7	0.53	1.36	1.32	87.45	7.92	21.93	44
Ctenochaetus_binotatus	Acanthuridae	IWP	0.71	0.36	1	1.12	1.32	0.63	15.2	8.48	85.38	64
Pomacanthus_semicirculatus	Pomacanthidae	IWP	0.57	0.36	1	1.03	1.32	0.86	25.88	9.05	74.18	46

Table 3. *Cont.*

Species	Family/Sub-Fam	Range	AVERAGE ABUNDANCE			Diss/SD			Cum.Av. Diss.Contrib %			Rank AvDiss/SD
			COMORO	MAINLAND	MADAGA	Comoros-Mainland	Madagacar-Mainland	Madagascar-Comoros	Comoros-Mainland	Madagacar-Mainland	Madagascar-Comoros	
Ctenochaetus_truncatus	Acanthuridae	IO	0.86	0.79	0.3	0.64	1.24	1.32	79.28	10.15	20.29	48
Zebrasoma_velifer	Acanthuridae	IWP	0.71	0.36	0.9	1.12	1.24	0.69	14.68	10.69	80.58	69
Pomacentrus_caeruleus	Pomacentridae	IO [3]	0.71	0.39	1	1.08	1.22	0.63	17.28	11.23	88.06	>180
Pterocaesio_tile	Caesionidae	IWP	0.86	0.79	0.3	0.64	1.24	1.32	79	11.77	20.84	49
Calotomus_carolinus	Scarinae	IP	0.57	0.68	0.2	0.94	1.23	1.07	44.84	12.29	41.67	42
Additional Significant Species (Top 20) in Comoros–Mainland (Note *C. argus* and *A. allardi*, above, also Rank Here)												
Heniochus_acuminatus	Chaeotodontidae	IWP	0	0.79	0.6	1.88	0.88	1.2	1.45	60.56	27.75	12
Aprion_virescens	Lutjanidae	IWP	0.71	0.21	0.5	1.27	0.99	0.99	6.43	46.28	47.41	40
Ostracion_meleagris	Ostraciidae	IP	0.71	0.21	0.6	1.27	1.11	0.9	5.84	19.3	60.75	37
Chromis_lepidolepis	Pomacentridae	IWP	0.29	0.75	0.5	1.23	0.99	0.99	8.14	32.72	55.92	45
Chaetodon_interruptus	Chaeotodontidae	IO	0.71	0.29	0.7	1.19	1.17	0.83	10.37	14.36	72.76	50
Anampses_lineatus	Labridae	RS-IO	0.71	0.32	0.3	1.15	0.86	1.17	10.92	63.97	29.77	51

[1] In IO reported only from East Africa, Seychelles, Madagascar, Mascarenes, Maldives, Chagos Archipelago, Andaman Sea and Java (Fricke et al. 2021). [2] and Sri Lanka, RFM pers.obs. [3] WIO to western Indonesia, RFM pers.obs.

Legend:

significant in Mada-Com and Mada-Main	only significant in Mada-Com	significant in Mada-Com and Com-main	significant in Mada-main and Com-main	only significant in Mada-main	only significant in Com-main

The mean Dissimilarity/SD across all three pairwise comparisons provides a metric to rank species as most significant in all three SIMPER comparisons, with one representing the most highly ranked, *Plectropomus punctatus* (Table 3). The first 27 species (*Variola louti* is 27th) that ranked most highly out of the 356 observed species all appear in the Madagascar–Comoros comparisons, except for *Cephalopholis argus* and *Chaetodon falcula*, which explain differences between Madagascar and mainland countries, and *Heniochus acuminatus*, which explains differences between Comoros and mainland countries (Table 3).

Taking the top 20 species that most contributed to explaining differences (average cumulative dissimilarity of 12–14%) in species assemblages between the three geographic zones, the following broad patterns can be seen (Table 3):

- the five most significant species in explaining differences between Comoros and Madagascar were very common in Madagascar but not sighted in Comoros: *Plectropomus punctatus, Plectorhinchus gaterinus, Hipposcarus harid, Epibulis insidiator* and *Chaetodon vagabundus*. All were moderately common in mainland countries except *C. vagabundus*, which was rare;
- other species explained differences between Comoros and Madagascar but did not appear in the top 20 species in other paired geographic zone comparisons, such as *Pomacentrus baenschi* and *P. caeruleopunctatus*, or were species common in Madagascar (in 80% of sites) but either rare or absent in mainland;
- high ranking species in Comoros were not sighted in Madagascar, e.g., *Anampses twistii* and *Pygoplites diacanthus*. *Abudefduf vaigiensis*, though less significant, was also common in Comoros and not sighted in Madagascar. Other species common in Comoros but rare/absent in Madagascar included *Coris formosa, Mulloidichthys flavolineatus,* and *Parupeneus trifasciatus;*
- other species that contributed to both Madagascar—Comoros and Madagascar—mainland comparisons were common or relatively common across sites in Madagascar and mainland but were rare or absent in Comoros, such as *Neoglyphidodon melas, Scarus ghobban, Cheilinus fasciatus* and *Cheilinus oxycephalus,* and *Chaetodon melannotus*. With a similar distribution pattern though less significant was *Lutjanus fulviflamma* which was rare in Comoros. *Scarus ghobban,* (ranked 10th overall), was a key species distinguishing Comoros, where it was rare, from both Madagascar (8th rank) and mainland (20th rank, Table S3);
- species that were top ranking species contributing to the Comoros—mainland differences were not highly ranked in the other SIMPER results such as *Aprion virescens, Ostracion meleagris* and *Heniochus acuminatus*. The first two were more common in Comoros compared with mainland or Madagascar, whereas *H. acuminatus* was not sighted in Comoros. *Chaetodon interruptus* and *Anampses lineatus* were also common in Comoros and less common in mainland though were less significant in explaining differences;
- *Amphiprion allardi* and *Cephalopholis argus* were the only species significant in both island—mainland comparisons: *A. allardi* was only observed in mainland sites and *C. argus* was much more common in mainland sites.

In summary, the top 22 species that contributed most significantly to the differences in species occurrence and assemblage homogeneity between the three geographic zones (Table 3) were from 10 of the 19 families surveyed (where Scarinae are separated from Labridae): Labridae (six species), Chaetodontidae (4) Pomacentridae (3), Scarinae and Epinephelidae (two species of each), and one species in each of Acanthuridae, Lutjanidae, Haemulidae, Pomacanthidae, and Ostraciidae. Notably, very few larger bodied fishery species ranked highly in the dissimilarity rankings except *Pletropomus punctatus* and *Plectorhinchus gaterinus*. Five of the ten families ranked in the present study correspond to those most recommended as good indicators of reef fish diversity in other studies (Table 1). Notably absent from most of these other studies' recommendations are the Epinephelidae, Haemulidae and Ostraciidae.

4. Discussion

4.1. Biogeographic Patterns in Species Diversity

The designation of the WIO as one biogeographic marine Province [3,15,16] is supported by the species richness of coral reef fishes which was similar across the central WIO region surveyed, at ~55% similarity. However, within the study area, significant differences in species presence were found which separated locations in Comoros from northeast Madagascar and these both differed from mainland eastern Africa. The highest species richness was found in locations in Tanzania and northern Mozambique, with the lowest species richness and most different species composition in Comoros. A lower species richness of corals from Comoros is also reported [21]. These findings, based on a sample of 356 species, provide evidence for three ecoregions within this central region of the WIO: eastern Africa (Tanzania, including Zanzibar, and northern Mozambique); Comoros; and north-eastern Madagascar. These ecoregions can represent conservation planning units for Marine Spatial Planning [45] or the Red Listing of Ecosystems [46].

Biogeography is a primary driver of patterns in species richness of reef fishes [1,3,15,47]. Factors such as the island effect, reef area, coast length and sea surface temperature are significant elements of this biogeography [5,19,48,49]. The three islands of the Union of Comoros are small, oceanic and recent volcanic islands [50] isolated from other large coral reef systems: ~250 km from Mozambique and ~450 km from Madagascar. The total coastline length and reef area in Comoros are ~70–90% smaller than mainland east Africa or Madagascar [51]. These biogeographic factors therefore likely explain the naturally lower species richness and different assemblages in Comoros. Less diversity of habitats may also be a contributing factor, but this could not be tested with the current sampling design. However, taxonomic diversity was highest in Comoros, driven by the Tetradontiformes. Likely reasons can only be speculative currently, but they may relate to the unique steep bathymetry of these volcanic islands.

Present day ocean currents [19] that drive the dispersal of pelagic larvae also help maintain biogeographic patterns. Further evidence for the ecoregions of Comoros, northern Madagascar and mainland east Africa comes from modelling pelagic larval duration (PLD) which separated eastern Africa from Comoros and Northern Madagascar using a short PLD of 10 days [52], generally shorter than most reef fishes. However, at 50 PLD this area became one homogeneous region, a finding verified by the genetics of two species common in the present study, *Epinephelus merra* and *Lutjanus kasmira* [53,54]. This finding lends further weight to the delineation of the WIO as one biogeographic province.

4.2. Anthropogenic Influences

Reefs of the WIO are intensely fished in many locations [38,55,56], climate induced coral bleaching continues to degrade reefs [57] and both these impacts threaten some reef fish species [9] which could undermine studies examining patterns in species richness. However, fishing effects are largely seen in declines in biomass and fish size data [51,58,59], so with the species presence/absence data in the current study fishing effects will only manifest in zero values. Five species completely absent from Comoros and in the top 12 ranking species in the SIMPER analyses are species typically taken in WIO artisanal fisheries [38,60]. Two of these species, the grouper *Plectropomus punctatus* and the sweetlip *Plectorhinchus gaterinus*, are reported historically in Comoros [61] and at neighboring Mayotte the easternmost island in the Comoros Archipelago [62]. Both are widespread western Indian Ocean species and would be vulnerable to the coastal handline fisheries in Comoros [29] so their lack of sightings may indicate fishing effects. No confirmed sightings of *Plectropomus punctatus* in subsequent surveys in Moheli island in 2016 (B. Cowburn pers. comm.) and 2018 (M. Samoilys pers. obs.) further confirm this species is now very rare or locally extinct in these islands. *Plectorhinchus gaterinus* was also not sighted in surveys in Comoros in 2016 or 2018 and is reported as locally extinct in Reunion [32]. These two fishery species are therefore likely to have regional distribution patterns skewed by fishing. The other fishery species absent in Comoros are more easily explained through range restrictions or habitat requirements (see below).

4.3. Species Level Differences

Knowledge of the diversity and distribution of coral reef fishes is important for conservation planning under future climate change scenarios [63], yet species level data are still lacking over large spatial scales [1]. The present study fills this gap revealing regional species distribution patterns with assemblages in Comoros less speciose compared with the other three countries. Several species not sighted in Comoros included wide ranging Indo–Pacific species, those widely reported in the WIO, as well as WIO endemics. For example, widely distributed Indo–Pacific species were *Epibulus insidiator*, *Chaetodon vagabundus*, and *Cheilinus fasciatus*. However, they were sighted in Comoros in subsequent surveys in 2016 (B. Cowburn pers. comm.) and *C. vagabundus* was also sighted in 2018 (M. Samoilys pers. obs.), suggesting the zero sightings in the current study reflect rarity not absence. The WIO endemic anemone fish *Amphiprion allardi* ranked highly in the SIMPER analysis because it is restricted to the mainland east African coast and is replaced by *A. latifasciatus* in Comoros and Madagascar [64]. Therefore, earlier records of *A. allardi* from Mayotte, Glorieuses and the Mascarene Islands [62,65,66] need updating. Other species not sighted in Comoros, *Hipposcarus harid* and *Lutjanus fulviflamma*, may be explained by restricted larval supply due to currents or habitat requirements. For example *H. harid* is reported from East Africa, western Madagascar [27], Mayotte island [62], and the French territories of Glorieuses and Geyser reef [67], but there are no records for Comoros in the Catalog of Fishes [42]. The reasons for this apparent disjunct distribution in the Comoros Archipelago require further study. *Lutjanus fulviflamma* is widely distributed in the WIO including Mayotte island [62]. Its absence in Comoros may be due to the juveniles' strong dependence on mangroves, which are uncommon in the islands of the Union of Comoros [51]. Only two species were common to Comoros but uncommon in the mainland countries and only moderately common in Madagascar: the snapper *Aprion virescens* and the boxfish *Ostracion meleagris*. This is possibly related to the predominantly steep narrow forereefs of these volcanic islands.

Principle species with restricted ranges that were driving regional differences include two Pomacentridae, *Pomacentrus baenschi* and *Pomacentrus caeruleopuntatus*, both reported as restricted to Madagascar and the Mascarene Plateau [65,68]. In the present study they were common in northeast Madagascar but sighted, though uncommon, in mainland (both species) and Comoros (*P. caeruleopuntatus*). These sightings require further investigation. In contrast, other species were ubiquitous in some locations and not in others, for example the parrotfish *Scarus ghobban*, with a wide Red Sea—Indo–West Pacific range [33], was sighted at every site in Madagascar, at just over half the mainland sites, but only 14% of the sites in Comoros. This pattern may reflect the small, isolated island effect (see above) and less diversity of habitats in Comoros.

Species missing from the northeast Madagascar sites are possibly explained by range restrictions due to ocean currents. *Pygoplites diacanthus* and *Anampses twistii* were not seen and *Cephalopholis argus* and *Chaetodon falcula* were rare, yet all are reported as widely distributed in the Indo–West Pacific [42] and in northwest Madagascar [27]. The South Equatorial Current (SEC) bifurcates east of Madagascar with the northern current flowing over the northern tip of Madagascar to continue to the Comoros and then the African continent, while the southern flow travels down the east coast of Madagascar to join the Algulhas current off South Africa [69]. The northeastern location of the present study may therefore represent Madagascan reefs where larval supply is weak because the SEC bifurcates further east, so self-recruitment is more prevalent [52]. Larval connectivity between the SEC, the west coast of Madagascar and the Comoros is likely to be strong due to the gyres in the north of the Mozambique Channel around the Comoros Archipelago [70]. Thus, species absent in northeast Madagascar are more likely explained by ocean currents restricting larval supply rather than "near threatened" as reported for Reunion where there are greater human pressures [65].

The ranges of the 46 species that contributed most to the patterns in diversity of species across this central WIO region were largely (65%) highly wide ranging (Indo–Pacific, Red Sea—Indo–West Pacific) [42]. A further 30% were wide ranging within the Indian Ocean, including the WIO, or Red Sea—Indian Ocean [15]. Only 4.3% of species were highly restricted, to the Mascarene Plateau [27]. Of conservation interest are the two nominally Mascarene species (*Pomacentrus caeruleopuntatus* and *P. baenschi*) and the six species restricted to the WIO Province, two of which (*Plectropomus punctatus* and *Plectorhinchus gaterinus*) are in severe decline [51]. Species that are endemic to the WIO [71], yet not seen during these surveys, provide a list of species of potential concern that may be disappearing due to loss of habitats through coral mortality. Additional surveys are recommended for these species, to be considered for Red Listing by the Species Survival Commission [72]. One example is *Chaetodon blackburni*, last assessed in 2010 as Least Concern [72], only know to occur in East and southern Africa, from Kenya to 33° S, and Madagascar & Mauritius [42], but not sighted once in this study. The functional roles of reef fish species, their contributions to ecosystem processes, are never equal, and it is postulated that in tropical systems each species contributes relatively little compared with temperate systems, due to high diversity in the tropics [73]. Further study on the ecological role of the 46 most significant species driving the regional patterns in the current study may reveal functional attributes important in ecosystem processes on coral reefs.

4.4. Methods for Species Richness Surveys

The presence/absence of species from 19 families surveyed by the timed UVC method in this study recorded ~45% of the total putative number of reef species in the WIO [19] and was effective in detecting biogeographical patterns in assemblages and significant differences in species richness between locations. Further, total number of species per location were not different from predicted values, giving validity to the method and the total numbers of species observed. The biogeographical patterns span the central WIO across a latitudinal and longitudinal range of around eight and eleven degrees, respectively. Taxonomic distinctness also differed across this region suggesting it was valuable adding the Tetradontiformes order (Trigger, Box, Puffer and File fish families) to the check list, in addition to the more commonly monitored Perciform families. These values and diversity patterns provide a benchmark for species assemblages prior to the 2016 mass coral bleaching event, which was widespread in the WIO causing significant coral mortality [74].

Ten of the nineteen families contributed the top 22 species that showed significant differences in species occurrence and assemblage homogeneity between the three geographic zones. Of these, the highest number of species were in the Labridae, Chaetodontidae, Pomacentridae, Scarinae and Epinephelidae. These match four of the six families in the widely used Coral Fish Diversity Index (CFDI) [75]. The other five significant families in the present study were Acanthuridae, Lutjanidae, Pomacanthidae, Haemulidae and Ostraciidae. Notably, many of the families important in local fisheries, such as Lethrinidae, Siganidae, Carangidae, did not rank, even in the top 46 species significant in explaining location differences. For rapid assessment surveys that have to address multiple issues and also need to collect density and fish size estimates, a reduced list of eight families would suffice, such as: Chaetodontidae, Scarinae, Epinephelidae, Acanthuridae, Lutjanidae, Pomacanthidae, Haemulidae and Ostraciidae. These families were significant in the present study and are either speciose, good reef fish indicators, or have fishery importance [37,39,76–78]. Pomacentridae and Labridae are important highly speciose families, but being small and often difficult to identify are often not counted to species level in standard monitoring programs, such as GCRMN [79]. However, they added valuable diversity data here and are recommended if species identifications are possible. For detailed conservation planning, species level data are preferable.

5. Conclusions

This study highlights that timed SCUBA surveys of reef fish species presence provide diversity metrics that are sensitive to change and can be used for conservation planning and to detect future impacts of conservation or reef degradation. For example, results indicate that conservation action in Comoros should prioritize protection of *Plectropomus punctatus* and *Plectorhynchus gaterinus*. The 22 most significant species that revealed patterns in diversity across the region came from the Scarinae, Chaetodontidae, Pomacentridae, Epinephelidae, Acanthuridae, Lutjanidae, Pomacanthidae, Haemulidae, and Ostraciidae suggesting these families should be considered for UVC surveys of reef fishes aiming to examine management, fishing effects and climate change on coral reefs. The current study provides a useful reference point for testing predictions of changes in reef fish species richness due to warming seas [63,80].

The study documents the occurrence of 356 species of reef associated fishes for locations in four countries in the WIO that lie across the northern Mozambique Channel area where the highest coral diversity in the WIO is found [21]. Testing that this region is a diversity hotspot for reef fish species will require comparable data from more peripheral sites such as in Kenya, Seychelles, Mauritius and Reunion. Our findings provide evidence that the WIO biogeographic province contains distinguished ecoregions: Comoros Archipelago; northeastern Madagascar; and northern Mozambique and Tanzania, including Zanzibar. This separation of Comoros from eastern Africa differs from the ecoregions based on hermatypic corals [22,24]. These intraregional differences are relevant in coral reef threat assessments and informed the recent IUCN Red Listing of Ecosystems process for the WIO's coral reefs [81].

Supplementary Materials: The following are available online at https://www.mdpi.com/article/10.3390/d14020102/s1, Figure S1: Total number of species per survey site (n = 45) in >75 min of observations, Figure S2: MDS plot showing four distinct groupings of the sites in terms of their species richness with Comoros and Madagascar separated and different from each other and the two mainland countries, Tanzania and Mozambique grouped together, with two outlier sites; Table S1: Final list of 45 survey sites used in analyses, showing countries, locations, reef geomorphology and reef type; Table S2: Full species inventory from all sites with taxonomic authority and taxonomic references. Table S3: Full SIMPER for all 356 species showing average Abundance, Dissimilarity, Cumulative contributions to Dissimilarity, and rankings.

Author Contributions: Conceptualization, M.S.; data curation, M.S. and L.A.-F.; formal analysis, L.A.-F.; writing—original draft, M.S.; writing—review & editing, M.S., L.A.-F., P.C. and R.M. All authors have read and agreed to the published version of the manuscript.

Funding: This research was funded by the Western Indian Ocean Marine Science Association (WIOMSA), MASMA grant OR/2008/05; and Conservation International for the Madagascar surveys. M.S. was partly funded by a grant from the Perivoli Trust for a Senior Research Fellowship at the University of Oxford.

Institutional Review Board Statement: Not applicable. All data are underwater visual observations.

Informed Consent Statement: Not applicable.

Data Availability Statement: Data from this study are provided in Tables S2 and S3 and will be made available on Dryad.

Acknowledgments: Grateful thanks to the following for comments and suggestions on the survey method: Howard Choat, Phil Heemstra, Ali Green, and Dave Bellwood. Gerry Allen's rapid reef fish diversity survey method for Conservation International, his comprehensive species lists for Madagascar, and help with identification of species from photographs were all invaluable. Thanks to Bemafaly Randriamanantsoa, Saleh Yahya, and Isabel da Silva as dive partners and for sharing their local knowledge of fish species. Thanks and appreciation to David Obura for stimulating discussions, fund raising, and help in several field trips.

Conflicts of Interest: The authors declare no conflict of interest. The funders had no role in the design of the study; in the collection, analyses, or interpretation of data; in the writing of the manuscript, or in the decision to publish the results.

References

1. Mora, C. Large-scale patterns and processes in reef fish richness. In *Ecology of Fishes on Coral Reefs*; Mora, C., Ed.; Cambridge University Press: Cambridge, UK, 2015; pp. 88–96.
2. Hubert, N.; Dettai, A.; Pruvost, P.; Cruaud, C.; Kulbicki, M.; Myers, R.F.; Borsa, P. Geography and life history traits account for the accumulation of cryptic diversity among Indo-West Pacific coral reef fishes. *Mar. Ecol. Prog. Ser.* **2017**, *583*, 179–193. [CrossRef]
3. Bellwood, D.R.; Wainright, P.C. The history and biogeography of fishes on coral reefs. In *Coral Reef Fishes*; Sale, P., Ed.; Academic Press: Cambridge, MA, USA, 2002; pp. 5–32.
4. Stow, D. *Vanished Ocean: How Tethys Reshaped the World*; Oxford University Press: Oxford, UK, 2010.
5. Parravicini, V.; Kulbicki, M.; Bellwood, D.R.; Friedlander, A.M.; Arias-Gonzalez, J.E.; Chabanet, P.; Floeter, S.; Myers, R.; Vigliola, L.; D'Agata, S.; et al. Global patterns and predictors of tropical reef fish species richness. *Ecography* **2013**, *36*, 1254–1262. [CrossRef]
6. Heron, S.F.; Maynard, J.A.; Van Hooidonk, R.; Eakin, C.M. Warming Trends and Bleaching Stress of the World's Coral Reefs 1985–2012. *Sci. Rep.* **2016**, *6*, 38402. [CrossRef] [PubMed]
7. Robinson, J.P.W.; Wilson, S.; Jennings, S.; Graham, N.A. Thermal stress induces persistently altered coral reef fish assemblages. *Glob. Chang. Biol.* **2019**, *25*, 2739–2750. [CrossRef] [PubMed]
8. Hughes, T.P.; Kerry, J.T.; Baird, A.H.; Connolly, S.R.; Dietzel, A.; Eakin, C.M.; Heron, S.; Hoey, A.S.; Hoogenboom, M.O.; Liu, G.; et al. Global warming transforms coral reef assemblages. *Nature* **2018**, *556*, 492–496. [CrossRef]
9. Graham, N.A.J.; Chabanet, P.; Evans, R.D.; Jennings, S.; Letourneur, Y.; Macneil, M.A.; McClanahan, T.R.; Öhman, M.C.; Polunin, N.V.C.; Wilson, S.K. Extinction vulnerability of coral reef fishes. *Ecol. Lett.* **2011**, *14*, 341–348. [CrossRef]
10. Graham, N.A.J.; McClanahan, T.R.; MacNeil, M.A.; Wilson, S.K.; Cinner, J.E.; Huchery, C.; Holmes, T.H. Human Disruption of Coral Reef Trophic Structure. *Curr. Biol.* **2017**, *27*, 231–236. [CrossRef]
11. Dana, J.D. On an isothermal oceanic chart illustrating the geographical distribution of marine animals. *Am. J. Sci.* **1853**, *16*, 153–167, 314–327.
12. Forbes, E. *The Natural History of the European Seas*; Goodwin-Austin, R., Ed.; Facsimile Publisher: London, UK, 1859.
13. Briggs, J.C. *Marine Zoogeography*; McGraw-Hill: New York, NY, USA, 1974.
14. Ekman, S. *Zoogeography of the Sea*; Sidgwick and Jackson: London, UK, 1953.
15. Briggs, J.C.; Bowen, B.W. A realignment of marine biogeographic provinces with particular reference to fish distributions. *J. Biogeogr.* **2012**, *39*, 12–30. [CrossRef]
16. Spalding, M.D.; Fox, H.E.; Allen, G.R.; Davidson, N.; Ferdaña, Z.A.; Finlayson, M.; Halpern, B.S.; Jorge, M.A.; Lombana, A.; Lourie, S.A.; et al. Marine ecoregions of the world: A bioregionalization of coastal and shelf areas. *Bioscience* **2007**, *57*, 573–583. [CrossRef]
17. IUCN. *Managing Marine Protected Areas: A Toolkit for the Western Indian Ocean*; IUCN Eastern Africa Regional Programme: Nairobi, Kenya, 2004.
18. McClanahan, T.R.; Sheppard, C.R.C.; Obura, D.O. (Eds.) *Coral Reefs of the Indian Ocean*; Oxford University Press: Oxford, UK, 2000.
19. Barber, P.H.; Meyer, C.P. Pluralism explains diversity in the Coral Triangle. In *Ecology of Fishes on Coral Reefs*; Mora, C., Ed.; Cambridge University Press: Cambridge, UK, 2015; pp. 258–263.
20. Carpenter, K.E.; Springer, V.G. The center of the center of marine shore fish biodiversity: The Philippine Islands. *Environ. Biol. Fishes.* **2005**, *72*, 467–480. [CrossRef]
21. Obura, D.O. An Indian Ocean centre of origin revisited: Palaeogene and Neogene influences defining a biogeographic realm. *J. Biogeogr.* **2015**, *43*, 229–242. [CrossRef]
22. Obura, D.O. The Diversity and Biogeography of Western Indian Ocean Reef-Building Corals. *PLoS ONE* **2012**, *7*, e45013. [CrossRef] [PubMed]
23. Obura, D.O.; Church, J.E.; Gabrié, C. *Assessing Marine World Heritage from an Ecosystem Perspective: The Western Indian Ocean*; UNESCO World Heritage Centre: Paris, France, 2012.
24. Veron, J.; Stafford-Smith, M.; DeVantier, L.; Turak, E. Overview of distribution patterns of zooxanthellate Scleractinia. *Front. Mar. Sci.* **2015**, *2*, 81. [CrossRef]
25. Burgess, N.D.; DAHales, J.D.A.; Underwood, E.; Dinerstein, E. (Eds.) *Terrestrial Ecoregions of Africa and Madagascar: A Conservation Assessment*; Island Press: Washington, DC, USA, 2004.
26. Mora, C.; Chittaro, P.M.; Sale, P.F.; Kritzer, J.P.; Ludsin, S.A. Patterns and processes in reef fish diversity. *Nature* **2003**, *421*, 933–936. [CrossRef]
27. Allen, G.R. Reef fishes of Northwestern Madagascar. In *A Rapid Marine Biodiversity Assessment of the Coral Reefs of Northwest Madagascar*; McKenna, S., Allen, G.R., Eds.; RAP Bullet; Conservation International: Washington, DC, USA, 2005; pp. 39–48.

28. Samoilys, M.; Randriamanantsoa, B. Reef fishes of northeast Madagascar. In *A Rapid Marine Biodiversity Assessment of the Coral Reefs of Northeast Madagascar*; Obura, D., Di Carlo, G., Rabearisoa, A., Oliver, T., Eds.; Conservation International: Washington, DC, USA, 2011; pp. 29–39.

29. Cowburn, B.; Samoilys, M.A.; Obura, D. The current status of coral reefs and their vulnerability to climate change and multiple human stresses in the Comoros Archipelago, Western Indian Ocean. *Mar. Pollut. Bull.* **2018**, *133*, 956–969. [CrossRef]

30. Fricke, R.; Durville, P.; Bernardi, G.; Borsa, P.; Mou-Tham, G.; Chabanet, P. Checklist of the shore fishes of Europa Island, Mozambique Channel, southwestern Indian Ocean, including 302 new records. *Stuttgarter Beiträge zur Naturkd A Neue Ser.* **2013**, *6*, 247–276.

31. Chabanet, P.; Durville, P. Reef Fish Inventory of Juan De Nova's Natural Park (Western Indian Ocean). *West Indian Ocean J. Mar. Sci.* **2005**, *4*, 145–162. [CrossRef]

32. Fricke, R. *Fishes of the Mascarene Islands (Mauritius, Réunion, Rodriguez). An Annotated Checklist, with Descriptions of New Species*; Theses Zoo; Koeltz Scientific Books: Königstein, Germany, 1999.

33. Lieske, E.; Myers, R.F. *Coral Reef Fishes*, 2nd ed.; Princeton University Press: Princeton, NJ, USA, 2001.

34. Smith, M.; Heemstra, P.C. *Smith's Sea Fishes*; Springer-Verlag: Berlin, Germany, 1996.

35. Nelson, J.S. *Fishes of the World*, 4th ed.; John Wiley & Sons Inc.: Hoboken, NJ, USA, 2006.

36. Kulbicki, M.; Vigliola, L.; Wantiez, L.; Floeter, S.; Hubert, N. The biogeography of Chaetodontidae, a model for reef fishes? In *The Biology and Ecology of Butterfly-Fishes*; Pratchett, M., Berumen, M., Kapoor, B., Eds.; CRC Press: Boca Raton, FL, USA, 2013; pp. 70–106.

37. Samoilys, M.; Roche, R.; Koldewey, H.; Turner, J. Patterns in reef fish assemblages: Insights from the Chagos Archipelago. *PLoS ONE* **2018**, *13*, e0191448. [CrossRef]

38. Samoilys, M.A.; Osuka, K.; Maina, G.W.; Obura, D.O. Artisanal fisheries on Kenya's coral reefs: Decadal trends reveal management needs. *Fish Res.* **2017**, *186*, 177–191. [CrossRef]

39. Bellwood, D.R. The Eocene fishes of Monte Bolca: The earliest coral reef fish assemblage. *Coral Reefs.* **1996**, *15*, 11–19. [CrossRef]

40. Choat, J.H.; Bellwood, D.R. Reef fishes: Their history and evolution. In *The Ecology of Fishes on Coral Reefs*; Sale, P.F., Ed.; Academic Press, Inc.: San Diego, CA, USA, 1991; pp. 39–61.

41. Munday, P.L.; Jones, G.P.; Pratchett, M.S.; Williams, A.J. Climate change and the future for coral reef fishes. *Fish Fish* **2008**, *9*, 261–285. [CrossRef]

42. Fricke, R.; Eschmeyer, W.N.; van der Laan, R. *Eschmeyer's Catalog of Fishes: Genera, Species, References*; Fricke, R., Eschmeyer, W.N., van der Laan, R., Eds.; California Academy of Sciences: San Francisco, CA, USA, 2021.

43. Clarke, K.R.; Warwick, R.M. A further biodiversity index applicable to species lists: Variation in taxonomic distinctness. *Mar. Ecol. Prog. Ser.* **2001**, *216*, 265–278. [CrossRef]

44. Warwick, R.M.; Clarke, K.R. Taxonomic Distinctness and Environmental Assessment. *J. Appl. Ecol.* **1998**, *35*, 532–543. [CrossRef]

45. Zauca, J.; Gee, K. *Maritime Spatial Planning: Past, Present, Future*; Zauca, J., Gee, K., Eds.; Springer: Heidelberg, Germany, 2018.

46. Keith, D.; Rodríguez, J.; Rodríguez-Clark, K.; Nicholson, E.; Aapala, K.; Alonso, A.; Asmussen, M.; Bachman, S.; Basset, A.; Barrow, E.G.; et al. Scientific Foundations for an IUCN Red List of Ecosystems. *PLoS ONE* **2013**, *8*, e62111. [CrossRef] [PubMed]

47. D'Agata, S.; Mouillot, D.; Kulbicki, M.; Andréfouët, S.; Bellwood, D.R.; Cinner, J.E.; Cowman, P.; Kronen, M.; Pinca, S.; Vigliola, L. Human-mediated loss of phylogenetic and functional diversity in coral reef fishes. *Curr. Biol.* **2014**, *24*, 555–560. [CrossRef]

48. Barneche, D.R.; Rezende, E.L.; Parravicini, V.; Maire, E.; Edgar, G.J.; Stuart-Smith, R.D.; Arias-González, J.E.; Ferreira, C.E.L.; Friedlander, A.M.; Green, A.L.; et al. Body size, reef area and temperature predict global reef-fish species richness across spatial scales. *Glob. Ecol. Biogeogr.* **2019**, *28*, 315–327. [CrossRef]

49. Bellwood, D.R.; Hughes, T.P.; Connolly, S.R.; Tanner, J. Environmental and geometric constraints on Indo-Pacific coral reef biodiversity. *Ecol. Lett.* **2005**, *8*, 643–651. [CrossRef]

50. Audru, J.C.; Guennoc, P.; Thinon, I.; Abellard, O. Bathymay: La structure sous-marine de Mayotte révélée par l'imagerie multifaisceaux. *Comptes. Rendus. Geosci.* **2006**, *338*, 1240–1249. [CrossRef]

51. Samoilys, M.A.; Halford, A.; Osuka, K. Disentangling drivers of the abundance of coral reef fishes in the Western Indian Ocean. *Ecol. Evol.* **2019**, *9*, 4149–4167. [CrossRef]

52. Crochelet, E.; Roberts, J.; Lagabrielle, E.; Obura, D.; Petit, M.; Chabanet, P. A model-based assessment of reef larvae dispersal in the Western Indian Ocean reveals regional connectivity patterns—Potential implications for conservation policies. *Reg. Stud. Mar. Sci.* **2016**, *7*, 159–167. [CrossRef]

53. Muths, D.; Tessier, E.; Bourjea, J. Genetic structure of the reef grouper Epinephelus merra in the West Indian Ocean appears congruent with biogeographic and oceanographic boundaries. *Mar. Ecol.* **2015**, *36*, 447–461. [CrossRef]

54. Muths, D.; Gouws, G.; Mwale, M.; Tessier, E.; Bourjea, J. Genetic connectivity of the reef fish Lutjanus kasmira at the scale of the western Indian Ocean. *Can. J. Fish. Aquat. Sci.* **2012**, *69*, 842–853. [CrossRef]

55. Jacquet, J.; Fox, H.; Motta, H.; Ngusaru, A.; Zeller, D. Few data but many fish: Marine small-scale fisheries catches for Mozambique and Tanzania. *Afr. J. Mar. Sci.* **2017**, *32*, 2338. [CrossRef]

56. Rehren, J.; Samoilys, M.; Reuter, H.; Jiddawi, N. Integrating resource perception, ecological surveys, and fisheries statistics: A review of the fisheries in Zanzibar. *Rev. Fish Sci. Aquac.* **2020**, *30*, 1–18. [CrossRef]

57. Ateweberhan, M.; McClanahan, T.R.; Graham, N.A.J.; Sheppard, C.R.C. Episodic heterogeneous decline and recovery of coral cover in the Indian Ocean. *Coral Reefs.* **2011**, *30*, 739–752. [CrossRef]

58. DeMartini, E.E.; Friedlander, A.M.; Sandin, S.A.; Sala, E. Differences in fish-assemblage structure between fished and unfished atolls in the northern Line Islands, central Pacific. *Mar. Ecol. Prog. Ser.* **2008**, *365*, 199–215. [CrossRef]
59. DeMartini, E.E.; Smith, J.E. Effects of fishing on the fishes and habitats of coral reefs. In *Ecology of Fishes on Coral Reefs*; Mora, C., Ed.; Cambridge University Press: Cambridge, UK, 2015; pp. 135–144.
60. Samoilys, M.A.; Osuka, K.; Mussa, J.; Rosendo, S.; Riddell, M.; Diade, M.; Mbugua, J.; Kawaka, J.; Hill, N.A.O.; Koldewey, H. An integrated assessment of coastal fisheries in Mozambique for conservation planning. *Ocean Coast Manag.* **2019**, *182*, 104924. [CrossRef]
61. Heemstra, P.C.; Randall, J.E. *Groupers of the World (Family Serranidae, Subfamily Epinephelinae). An Annotated and Illsutrated Catalogue of the Grouper*; Fisheries; FAO: Rome, Italy, 1993.
62. Wickel, J.; Jamon, A.; Pinault, M. *Composition et Structure des Peuplements Ichtyologiques Marins de l ' île de Mayotte (Sud-Ouest de L 'océan Indien)*; Société Française d'Ichtyologie: Paris, France, 2014; Volume 38, pp. 179–203.
63. Strona, G.; Lafferty, K.D.; Fattorini, S.; Beck, P.S.A.; Guilhaumon, F.; Arrigoni, R.; Montano, S.; Seveso, D.; Galli, P.; Planes, S.; et al. Global tropical reef fish richness could decline by around half if corals are lost. *Proc. R Soc. B Biol. Sci.* **2021**, *288*, 20210274. [CrossRef]
64. Fautin, D.G.; Allen, G.R. *Field Guide to Anemonefishes and their Host Sea Anemones*; West Australian Museum: Perth, Australia, 1992.
65. Fricke, R.; Mulochau, T.; Durville, P.; Chabanet, P.; Tessier, E.; Letourner, Y. Annotated checklist of the fish species (Pisces) of La Réunion, including a Red List of threatened and declining species. *Stuttgarter Beiträge zur Naturkd A Neue Ser.* **2009**, *2*, 1–168.
66. Durville, P.; Chabanet, P.; Quod, J.P. Visual Census of the Reef Fishes in the Natural Reserve of the Glorieuses Islands (Western Indian Ocean). *West. Indian Ocean J. Mar. Sci.* **2003**, *2*, 95–104. [CrossRef]
67. Chabanet, P.; Tessier, E.; Mulochau, T.; Durville, P.; Rene, F. Peuplement ichtyologique des Bancs des Geyser et Zelee (Ocean Indien Occidental). *Cybium* **2002**, *26*, 11–26.
68. Bourjon, P.; Crochelet, E.; Fricke, R. First record of the large caerulean damselfish, *Pomacentrus caeruleopunctatus* (Actinopterygii: Perciformes: Pomacentridae), from reunion island, south-west indian ocean. *Acta Ichthyol. Piscat.* **2019**, *49*, 59–63. [CrossRef]
69. Schott, F.A.; Xie, S.P.; McCreary, J.P. Indian ocean circulation and climate variability. *Rev. Geophys.* **2009**, *47*, 1–46. [CrossRef]
70. Lutjeharms, J.R.E.; Bornman, T.G. The importance of the greater agulhas currentis increasingly being recognized. *S. Afr. J. Sci.* **2010**, *106*, 1–4. [CrossRef]
71. Kulbicki, M.; Parravicini, V.; Bellwood, D.R.; Arias-Gonzàlez, E.; Chabanet, P.; Floeter, S.R.; Friedlander, A.; McPherson, J.; Myers, R.E.; Vigliola, L.; et al. Global biogeography of reef fishes: A hierarchical quantitative delineation of regions. *PLoS ONE* **2013**, *8*, e81847. [CrossRef]
72. IUCN. The IUCN Red List of Threatened Species. *Version 2021-2 [Internet]*. 2021. Available online: https://www.iucnredlist.org (accessed on 19 November 2021).
73. Stuart-Smith, R.D.; Bates, A.E.; Lefcheck, J.S.; Duffy, J.E.; Baker, S.C.; Thomson, R.J.; Stuart-Smith, J.F.; Hill, N.A.; Kininmonth, S.J.; Airoldi, L.; et al. Integrating abundance and functional traits reveals new global hotspots of fish diversity. Nature. *Nature* **2013**, *501*, 539–542. [CrossRef]
74. Gudka, M.; Obura, D.; Mbugua, J.; Ahamada, S.; Kloiber, U.; Holter, T. Participatory reporting of the 2016 bleaching event in the Western Indian Ocean. *Coral Reefs* **2020**, *39*, 1–11. [CrossRef]
75. Allen, G.R.; Werner, T.B. Coral reef fish assessment in the coral triangle of southeastern Asia. *Environ. Biol. Fishes* **2002**, *65*, 209–214. [CrossRef]
76. Cole, A.C.; Pratchett, M.S.; Jones, J.P. Diversity and functional importance of coral-feeding fishes on tropical coral reefs. *Fish Fish* **2008**, *9*, 286–307. [CrossRef]
77. Samoilys, M.A.; Carlos, G. Determining methods of underwater visual census for estimating the abundance of coral reef fishes. *Environ. Biol. Fishes* **2000**, *57*, 289–304. [CrossRef]
78. Pratchett, M.S.; Munday, P.L.; Wilson, S.K.; Graham, N.A.J.; Cinner, J.E.; Bellwood, D.R.; Jones, G.P.; Polunin, N.V.; Mcclanahan, T.R. Effects of climate-induced coral bleaching on coral-reef fishes ecological and economic consequences. *Oceanogr. Mar. Biol.* **2008**, *46*, 251–296. [CrossRef]
79. Obura, D.O.; Gudka, M.; Abdou Rabi, F.; Bacha Gian, S.; Bigot, L.; Bijoux, J.; Freed, S.; Maharavo, J.; Mwaura, J.; Porter, S.N.; et al. *Coral Reef Status Report for the Western Indian Ocean*; Global Coral Reef Monitoring Network (GCRMN); International Coral Reef Initiative (ICRI); Indian Ocean Commission: Port Louis, Mauritius, 2017.
80. Pecl, G.T.; Araújo, M.B.; Bell, J.D.; Blanchard, J.; Bonebrake, T.C.; Chen, I.C.; Clark, T.D.; Colwell, R.K.; Danielsen, F.; Evengård, B.; et al. Biodiversity redistribution under climate change: Impacts on ecosystems and human well-being. *Science* **2017**, *355*, eaai9214. [CrossRef] [PubMed]
81. Obura, D.O.; Gudka, M.; Samoilys, M.; Osuka, K.; Mbugua, J.; Keith, D.A.; Porter, S.; Roche, R.; van Hooidonk, R.; Ahamada, S.; et al. Vulnerability to collapse of coral reef ecosystems in the Western Indian Ocean. *Nat. Sustain.* **2021**. [CrossRef]

Article

Taxonomic Diversity of Decapod and Stomatopod Crustaceans Associated with Pocilloporid Corals in the Central Mexican Pacific

Arizbeth Alonso-Domínguez [1], Manuel Ayón-Parente [2], Michel E. Hendrickx [3], Eduardo Ríos-Jara [2], Ofelia Vargas-Ponce [4], María del Carmen Esqueda-González [2] and Fabián Alejandro Rodríguez-Zaragoza [2,*]

[1] Programa de Doctorado en Ciencias en Biosistemática, Ecología y Manejo de Recursos Naturales y Agrícolas (BEMARENA), Centro Universitario de Ciencias Biológicas y Agropecuarias, Universidad de Guadalajara, Camino Ramón Padilla Sánchez No. 2100, Nextipac, Zapopan C.P. 45200, Mexico; arizbeth.alonso@alumnos.udg.mx

[2] Laboratorio de Ecología Molecular, Microbiología y Taxonomía, Departamento de Ecología, Centro Universitario de Ciencias Biológicas y Agropecuarias, Universidad de Guadalajara, Camino Ramón Padilla Sánchez No. 2100, Nextipac, Zapopan C.P. 45200, Mexico; manuel.ayon@academicos.udg.mx (M.A.-P.); eduardo.rios@academicos.udg.mx (E.R.-J.); carmen.esqueda@academicos.udg.mx (M.d.C.E.-G.)

[3] Laboratorio de Invertebrados Bentónicos, Unidad Académica Mazatlán, Instituto de Ciencias del Mar y Limnología, Unidad Mazatlán, Universidad Nacional Autónoma de México, Avenida Joel Montes Camarena S/N Apartado Postal 811, Mazatl C.P. 82040, Mexico; michel@ola.icmyl.unam.mx

[4] Departamento de Botánica y Zoología, Centro Universitario de Ciencias Biológicas y Agropecuarias, Universidad de Guadalajara, Camino Ramón Padilla Sánchez No. 2100, Nextipac, Zapopan C.P. 45200, Mexico; ofelia.vargas@academicos.udg.mx

* Correspondence: fabian.rzaragoza@academicos.udg.mx

Citation: Alonso-Domínguez, A.; Ayón-Parente, M.; Hendrickx, M.E.; Ríos-Jara, E.; Vargas-Ponce, O.; Esqueda-González, M.d.C.; Rodríguez-Zaragoza, F.A. Taxonomic Diversity of Decapod and Stomatopod Crustaceans Associated with Pocilloporid Corals in the Central Mexican Pacific. *Diversity* **2022**, *14*, 72. https://doi.org/10.3390/d14020072

Academic Editors: Michael Wink and Simone Montano

Received: 15 December 2021
Accepted: 18 January 2022
Published: 21 January 2022

Publisher's Note: MDPI stays neutral with regard to jurisdictional claims in published maps and institutional affiliations.

Abstract: Many crustacean species are obligate associates of pocilloporid corals, where they feed, reproduce, and find shelter. However, these coral-associated crustaceans have been poorly studied in the eastern tropical Pacific. Determining the crustacean richness and taxonomic distinctness could help in comparing different coral reefs and the potential effects of degradation. This study evaluated the spatio–temporal variation of the taxonomic diversity and distinctness of coral-associated crustaceans in four ecosystems of the Central Mexican Pacific (CMP) with different conditions and coral cover. In all ecosystems, 48 quadrants were sampled during the summer and winter for two years. A total of 12,647 individuals belonging to 88 species, 43 genera, and 21 families were recorded. The sampling effort yielded 79.6% of the expected species richness in the study area. Species rarity had 19% singletons, 4% doubletons, 22% unique, and 9% duplicate species; two species represented new records for the Mexican Pacific, and six were new to the CMP. This study recorded most of the symbiotic crustacean species in pocilloporid corals previously reported in the CMP. The taxonomic diversity and distinctness differed significantly between coral ecosystems and seasons, which was also visualized by nMDS ordination, showing an evident spatio–temporal variation in the taxonomic beta diversity.

Keywords: crustacean; coral-associated; western Mexico; *Pocillopora*; diversity

1. Introduction

Invertebrates are frequently associated with scleractinian corals of the genus *Pocillopora* [1]. Macrocrustaceans are the most representative coral-associated fauna. Among these, diverse assemblages find shelter among the *Pocillopora* branches [2,3]. Different taxa, including shrimps, crabs, isopods, and copepods, have been described as coral symbionts [4], presenting different degrees of specialization in form and function [5]. Several of these species are obligate symbionts, always and permanently associated with specific hosts, while other species are facultative symbionts that can also survive outside their host, usually on non-living substrates [1,5]. As a general trend, the coral-associated fauna depends on the host for feeding and refuge [6,7]. In addition, coral-associated crustaceans help to maintain coral health by performing cleaning activities, such as removing sediment and parasites [8,9]. Some species also have an active role in their defense against predators. For example, species of *Trapezia* defend the coral from predators, such as the crown star (*Acanthaster planci*) [10,11], and the shrimp *Alpheus lottini* protects the coral from coralivorous mollusks of the genus *Drupella* [12]. Crustaceans represent up to 80% of the coral-associated fauna [1,13], playing multiple ecological roles. They are part of different trophic relationships: acting as predators, parasites, herbivores, scavengers, and detritivores. Most obligate symbionts are mucus, suspension, or deposit feeders [1,5]. Thus, they link primary producers with high-level consumers [14–16]. This strong relationship between corals and crustaceans can be affected by the changes induced by anthropogenic activities or climate change [1,7].

The coral ecosystems of the Central Mexican Pacific (CMP) are dominated by several species of *Pocillopora* [17,18]. The most common is *P. verrucosa*, but other species, such as *P. damicornis*, *P. capitata*, *P. eydouxi*, *P. effusus*, *P. inflata*, and *P. meandrina*, have also been recorded [18]. Pocilloporid corals are structurally complex, generating many microhabitats for crustaceans [1]. However, few studies have focused on the crustacean diversity associated with pocilloporid corals in this area. Earlier studies in the Mexican Pacific by Pereyra-Ortega [19] and Hernández [20] described the decapods associated with *Pocillopora* corals in Isla Espíritu Santo and the southern area of the Baja California peninsula. Ramírez-Luna et al. [21] studied the temporal variation of the xanthid crabs in Huatulco Bay, Oaxaca, and found the largest diversity and abundance during the dry season. Hernández et al. [22,23] analyzed the impact of coral bleaching and hurricanes on the diversity and abundance of decapods from La Paz and Loreto Bay, Baja California Sur. They concluded that these phenomena changed the species richness considerably, decreasing the abundance of coral-associated decapod species. Two studies have evaluated the diversity of coral-associated crustaceans in the CMP, including the coastal region from Nayarit to Michoacan. Hernández et al. [24] performed a visual census of the decapods in coral ecosystems and found 36 species, with most individuals in or near corals. Ayón-Parente et al. [25] formulated an inventory of 19 species of caridean shrimps associated with the *Pocillopora* from Chamela Bay, Jalisco. Although both studies contributed to the inventories of the crustaceans associated with pocilloporid corals of the CMP, they did not offer evidence of possible spatio–temporal changes in their species richness and abundance, nor did they evaluate the contribution of the different taxonomic categories to diversity.

The average taxonomic distinctness (Δ^+) index has been used to assess biodiversity [26]. Environmental variability, sampling effort, and sampling size can affect most classical indices based on species richness and evenness [27]. However, the Δ^+ and its variation (Λ^+) are good ecological indicators, because they reflect the taxonomic relatedness of species within assemblages [28,29]. These indices allow for comparing different studies because they are independent of sample size and effort and provide a test for the significance of departure from expectation by chance if no other studies are available for comparison [30]. This analysis determines how certain taxa contribute to the total taxonomic diversity [26]. The taxonomic distinctness and its variation have mainly been used to evaluate biodiversity in time and space scales in different assemblages such as freshwater fishes [31,32], marine invertebrates [29,33,34], and insects [35,36].

In this study, the main objective was to use the species richness and taxonomic distinctness to assess the spatio–temporal variation of the decapods and stomatopods associated with the coral *Pocillopora* in the CMP. Knowing this information could help us understand the potential effects of coral reef degradation [7]. This area harbors the highest richness and coral coverage of the Mexican Pacific [17,18]; its coral ecosystems are dominated by the *Pocillopora* genus, which includes up to 80% of coral-associated fauna [1,13]. The CMP has suffered from a significant human impact and, although some areas are protected, most of the coral ecosystems are not [18]. Corals are very susceptible to environmental changes and natural and anthropogenic impacts. These changes affect associated fauna, especially symbiotic species. Evidence has shown that the *Pocillopora*-associated fauna has a spatio–temporal variation due to environmental drivers [3,6,21,37]. We hypothesized that the sites with the most discontinuous coral cover and highest human intervention (local tourism, fishing, etc.) would have the lowest richness and taxonomic distinctness, along with a high abundance of coral-associated fauna due to the low coverage and greater isolation of the host coral colonies.

2. Materials and Methods

2.1. Study Area

The study area included four coral ecosystems in the Central Mexican Pacific (CMP): (i) Chamela and (ii) Cuastecomate-Punta Melaque in southern Jalisco, and (iii) Carrizales and (iv) Punto B in Colima (Figure 1). The CMP is part of the eastern tropical Pacific ecoregion spanning from Baja California to northern Peru and the Galapagos Islands, Ecuador [38]. In the summer, the CMP is influenced by the California Current, the Cabo Corrientes Upwelling, the Mexican Warm Pool, and the Costa Rica Coastal Current. However, these currents have a weaker effect during the winter and spring due to the cold water from the California Current and the warm water from the Cortés Current [39]. Furthermore, the Mexican Warm Pool is part of the Western Hemisphere Warm Pool, which induces an important annual climatic variation in the water temperature to develop the El Niño-Southern Oscillation (ENSO) [40,41]. These currents provide the CMP with species from different biogeographic provinces [38]. Hurricanes, tropical storms, and upwellings also significantly impact the coral colony structure and associated fauna [18].

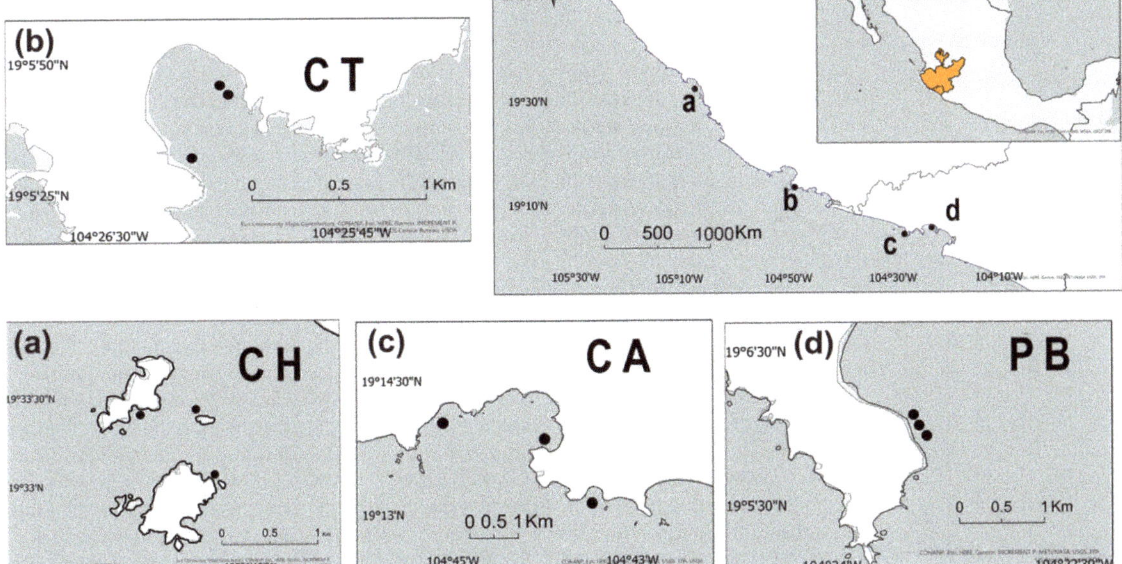

Figure 1. Study area in the CMP. Site codes: (**a**) CH, Chamela; (**b**) CT, Cuastecomate-Punta Melaque, Jalisco; (**c**) CA, Carrizales; and (**d**) PB, Punto B, Colima.

Some of the general characteristics of the sampled sites are as follows: (1) Chamela (CH) is formed by different small islands and islets; its coral ecosystems are patchy and isolated, and the benthos has a high coverage of rubble, sand, and dead coral. This site is important for fishing and local tourism. (2) Cuastecomate-Punta Melaque (CT) has a discontinuously distributed high coral cover, characterized by small reef patches with fleshy macroalgae stands, sand, and rocks. (3) Carrizales (CA) is located in Ceniceros Bay; it is a short beach defined by two small and fringing coral reefs on each side of the shore with ~100% live coral cover. (4) Punto B (PB) is located in Santiago Bay near the Julualpan Lagoon's mouth and is considered a highly touristic area. Its coral community has scarce, isolated coral colonies but great coverage of sponges, calcareous algae, and sandy and rocky substrates.

Three samples of live *Pocillopora* corals were collected in each coral ecosystem using randomly placed 0.25 m^2 quadrants. Each sample position was marked using a global positioning system (GPS). A total of 48 samples were collected during September 2017, January and September 2018, and January 2019. All samplings were obtained by scuba diving at a 10 m depth. Each coral sample was covered with a plastic bag to avoid losing organisms and detached using a hammer and chisel. Subsequently, the coral was carefully fragmented to collect all the organisms between and within the *Pocillopora* branches. All live decapods and stomatopods were fixed with 70% ethylic alcohol. Samples were identified to the most precise taxonomic level possible in the Molecular Ecology, Microbiology, and Taxonomy Laboratory (LEMITAX), Universidad de Guadalajara. The specialized literature for identification included Rathbun [42], Haig [43], Abele and Kim [44], Castro [45], Anker et al. [46], Hendrickx et al. [47], Ayón-Parente [48], García-Madrigal and Andréu-Sánchez [49], Hermoso-Salazar [50], Salgado-Barragán and Hendrickx [51], and Hiller and Lessios [52]. A presence/absence matrix was constructed to perform the ecological analysis.

2.2. Data Analysis

The spatial and temporal variation of the taxonomic diversity was evaluated with a three-way experimental design with crossed factors expressed as:

$$Y = \mu + Ye_i + Se_j + Si_k + Ye_i x Se_j + Ye_i x Si_k + Se_j x Si_k + Ye_i x Se_j x Si_k + \varepsilon_{ijk} \qquad (1)$$

where Y is the variable under analysis (taxonomic diversity), and μ is the mean of the analyzed variable. The year factor (Ye_i) had two levels (years), and each year was composed of two seasons (dry and wet seasons), so the first year included September 2017 and January 2018, and the second included September 2018 and January 2019. The season factor (Se_j) had two levels: wet (September 2017 and 2018) and dry (January 2018 and 2019). The site factor (Si_k) had four levels corresponding to the studied coral ecosystems. Finally, ε_{ijk} represented the accumulated error. All factors were considered as fixed effects (model type I).

The sampling effort was evaluated using sample-based rarefactions at three levels (i.e., site, season, and year) with the observed species richness and the expected richness estimated using non-parametric estimators Chao 1, Chao 2, Jackknife 1, Jackknife 2, ICE, and ACE. These estimators were based on rare species; they estimated the number of potential species considering the incidence and abundance data recorded in the samplings [53]. The total observed richness (S_{Obs}) was calculated for each ecosystem with the Mao Tao function. Then, the coral ecosystems were compared in pairs with individual-based rarefactions and 95% confidence intervals. All rarefaction curves were built with 10,000 randomizations without replacement. Species rarity was also calculated (singletons, doubletons, unique, and duplicate species), and the species were identified. These analyses were performed in the software EstimateS 9.1 [54]. The absolute density of each species (represented as the number of individuals per m^2) and their absolute frequency were also estimated.

The taxonomic diversity analysis considered each site's taxonomic differences and singularities regarding the seasonal variation and the years analyzed. Thus, the average taxonomic distinctness (Δ^+) analysis was performed to evaluate the species' distribution and incidence as well as their taxonomic relations [28]. This analysis also measured the taxonomic distance between two species and its variation (Λ^+), according to the following equations:

$$\Delta^+ = \left[\sum\sum_{i<j} \omega_{ij}\right] / [S(S-1)/2] \tag{2}$$

$$\Lambda^+ = \left[\sum\sum_{i \neq j} \omega_{ij-\omega} 2\right] / [S(S-1)/] \tag{3}$$

where S represents the number of species, and ω_{ij} denotes the assigned weight of each supraspecific taxonomic level. An eight-level taxonomic aggregation matrix was built, including species, genus, family, subfamily, suborder, order, subclass, and class. According to Warwick and Clarke [55], the taxa were weighted as follows: $\omega = 1$, species within the same genus; $\omega = 2$, species within the same family but different genus; $\omega = 3$, species within the same subfamily but in a different family; and so on. The Δ^+ and Λ^+ were estimated for each site, season, and year. The models were created with a 95% confidence interval, and the statistical significance was tested with 10,000 permutations.

The Δ^+ analysis was followed by a taxonomic dissimilarity analysis (Γ^+), which is described as:

$$\Gamma^+ = \frac{\left(\sum_{i=1}^{S_1} \min_j\{\omega_{ij}\} + \sum_{j=1}^{S_2} \min_i\{\omega_{ij}\}\right)}{(S_1 + S_2)} \tag{4}$$

where Γ^+ denotes the gamma$^+$ taxonomic dissimilarity, S_1 represents the number of species in the first sample, S_2 is the number of species in the second sample, and ω_{ij} denotes the path length between species i and j.

A non-parametric multidimensional scaling (nMDS) and a cluster analysis were performed using the taxonomic dissimilarity (Γ^+) matrix to explore the crustacean taxonomic differentiation patterns across the spatio–temporal experimental design (site, season, and year). The cluster analysis was built with the average group linking method and similarity profile analysis (SIMPROF) to assess group formation using 10,000 permutations. Therefore, nMDS ordination was coupled with the cluster analysis outputs. All analyses (i.e., Δ^+, Λ^+, Γ^+, nMDS, and cluster analysis) were performed in PRIMER 7.0.21 and PERMANOVA +1 [56].

3. Results

A total of 12,647 specimens were collected, representing 21 families, 43 genera, and 88 species (Supplementary Material, Table S1). For each quadrant, the number of species collected ranged from 13 to 38, and the number of individuals ranged from 36 to 705. The most diverse families were Alpheidae (21 species) and Porcellanidae (20 species). Ten families (47%) were represented by a single species (Supplementary Material, Table S1). The sample-based rarefaction showed that the sampling effort had an adequate representativity (79.6%) of the species richness expected by chance (Supplementary Material, Figure S1). The sampling effort ranged between 77.9% and 91.6% of representativity for all sites. The seasons showed 84.6% of representativity during the dry season and 85.7% in the wet season (Supplementary Material, Table S2). The representativity for year one was 79.5%, and for year two it was 88% (Supplementary Material, Table S2).

The most abundant species were *Trapezia corallina*, with 1720 individuals (13.6% of total abundance); *Trapezia bidentata*, with 1489; *Pachychelles biocellatus*, with 1028; *Petrolisthes haigae*, with 955; *Alpheus lottini*, with 820; *Petrolisthes hians*, with 619; and *Trapezia formosa*, with 579. Together, these species represented more than half of the collected specimens (Supplementary Material, Table S1). Of the total, 38 species (43% of the total) were represented by less than 10 individuals. Of these, 17 species had only 1 individual (singletons), and 4 had only 2 individuals (doubletons). Consequently, the contribution of singletons and doubletons to the species richness was 23.8%. In addition, 14 species were collected in only 1 sample (uniques) and 7 in 2 samples (duplicates) (Supplementary Material, Table S1).

Individual-based rarefactions in pairwise comparisons showed that the species richness between sites was similar because their confidence intervals (95%) overlapped (Supplementary Material, Figure S2). An exception was Chamela and Carrizales, which had the highest and lowest number of species, respectively (Supplementary Material, Figure S2). The highest total species richness and abundance recorded over the sampling period were as follows: for Chamela, 69 species and 2371 individuals; for Cuastecomate-Punta Melaque, 64 species and 3266 individuals; for Carrizales, 58 species and 2752 individuals; and for Punto B, 68 species and 4258 individuals (Supplementary Material, Table S1). The total species richness was similar between years and between seasons (Supplementary Material, Figure S3). Year one showed 78 species and 5957 individuals, and year two showed 76 species and 6690 individuals. The wet season showed 79 species and 5276 individuals, and the dry season showed 73 species and 7361 individuals (Supplementary Material, Table S1). *Synalpheus arostris* and *Neogonodactylus pumilus* were recorded for the first time in the Mexican Pacific and showed a geographic extension of 3950 km to the north. Six other species were recorded for the first time in the Central Mexican Pacific: *Lophopanopeus frontalis, Daldorfia trigona, Pilumnus gonzalensis, Pilumnus reticulatus, Tumidotheres margarita,* and *Megalobrachium tuberculipes*. In Chamela, 50% of the species were collected for the first time, in Cuastecomate-Punta Melaque, 76%, and in Carrizales and Punto B, 63%.

The average taxonomic distinctness (Δ^+) analysis at the site level showed that the Δ^+ values for all the sites fell inside the probability funnel or within the 95% confidence intervals ($p > 0.05$). Chamela had the lowest Δ^+ values despite having the greatest number of species (Figure 2). Punto B had the highest Δ^+ values above the global Δ^+ of the model. However, the Λ^+ values for all sites fell within the probability funnel, indicating that the sampled sites were representative of the taxonomic diversity of the area. The seasons had different Δ^+ values because the wet season fell within the probability funnel, but the dry season was outside the funnel ($p < 0.05$). The Λ^+ values for the dry season were outside the funnel, so the taxonomic representativity during the dry season was lower than expected by chance (Figure 2). The Δ^+ and Λ^+ values between years were similar and fell inside the probability funnel (Figure 2).

The nMDS ordination showed that the taxonomic dissimilarity (Γ^+) differed among the sites (Figure 3). The cluster analysis based on the SIMPROF procedure confirmed a group constituted by the southern sites (i.e., Carrizales and Punto B) and two separate entities (i.e., Chamela and Cuastecomate-Punta Melaque). This was also observed in the nMDS ordination. Carrizales and Punto B shared several species (e.g., *Pseudosquillisma adiastalta, Pomagnathus corallinus,* and *Synalpheus arostris*) and genera (e.g., *Trapezia, Liomera,* and *Pomaghnathus*), and they had almost the same families, except for the Pinnotheridae, which was only present in Punto B (and Cuastecomate-Punta Melaque). Conversely, Cuastecomate-Punta Melaque had a different crustacean fauna compared to the other sites and showed a mixture of taxa shared with Chamela and Punto B. Cuastecomate-Punta Melaque presented 34 genera and 17 families; these families were the same as Punto B, except for Panopeidae, which was found exclusively in this site, and Pseudosquillidae, which was absent. Chamela is the northernmost site and the most distant from the others. It was different because it had one superfamily (Parthenopoidea) not found elsewhere and two absent superfamilies (Eriphioidea and Pinnotheroidea). In Chamela, two families that were not found in other sites were collected (Parthenopidae and Lysmatidae), and four families (Panopeidae, Oziidae, Pseudosquillidae, and Pinnotheridae) were absent.

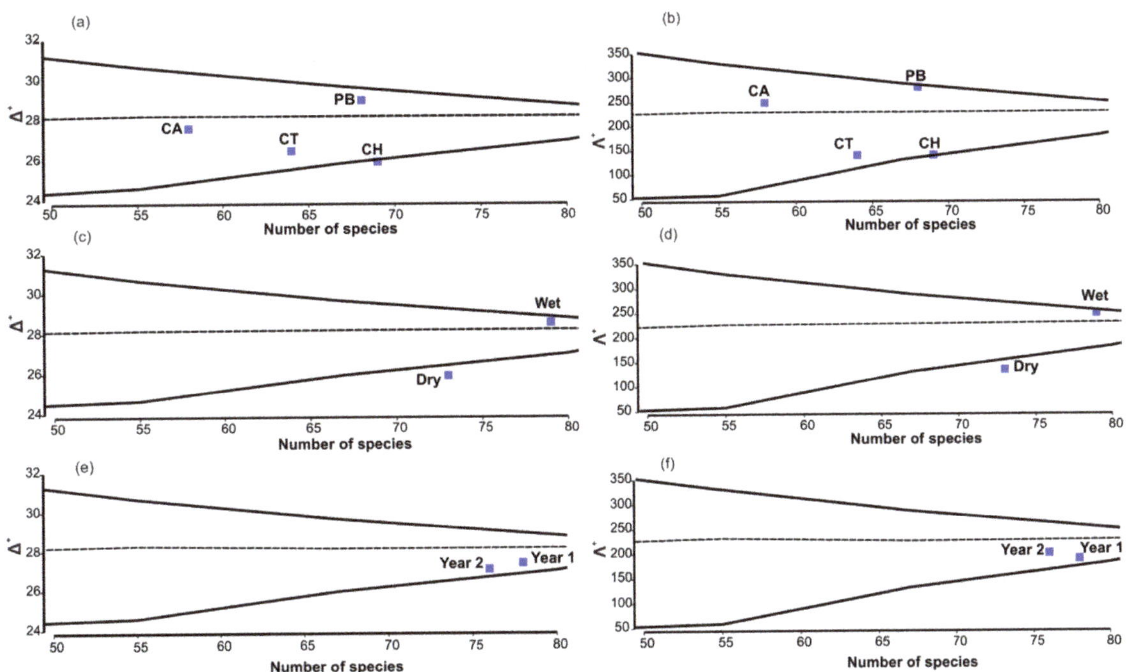

Figure 2. Average taxonomic distinctness analysis (Δ^+) by site (**a**), season (**c**), and year (**e**); and its variation Λ^+ for (**b**) site, (**d**) season, and (**f**) year. Codes: CH, Chamela; CT, Cuastecomate-Punta Melaque; CA, Carrizales; PB, Punto B.

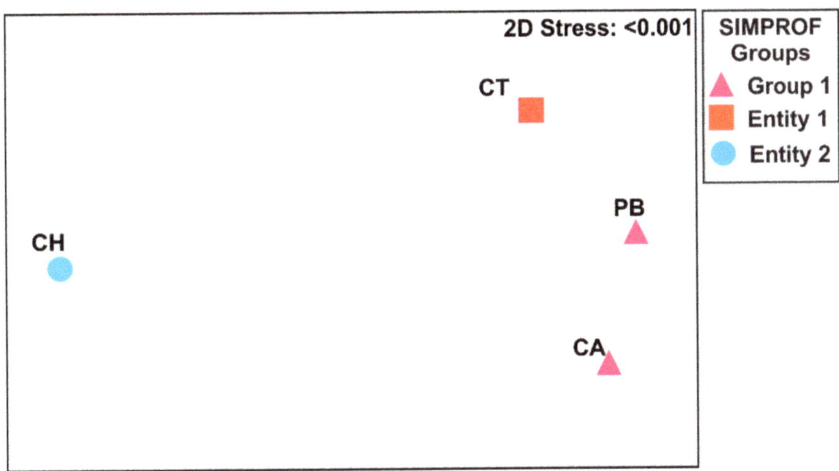

Figure 3. Non-metric multidimensional scaling (nMDS) ordination shows the taxonomic dissimilarity of the crustacean diversity associated with *Pocillopora* corals among the studied sites in the CMP. Groups were separated as a function of the cluster analysis with an average group linking method and the similarity profile analysis (SIMPROF). Codes: CH, Chamela; CT, Cuastecomate-Punta Melaque; CA, Carrizales; PB, Punto B.

4. Discussion

This study recorded most of the *Pocillopora* obligate symbiotic crustacean species reported by previous studies, including *Trapezia bidentata, T. corallina, T. digitalis, T. formosa, Alpheus lottini, Hapalocarcinus marsupialis,* and some species of *Synalpheus*. However, we did not find some species known to be associated with *Pocillopora*, such as *Fennera chacei, Alpheus sulcatus, Palaemonella holmesi, Stenorhynchus debilis, Thor algicola,* and *Petrolisthes galathinus,* which had been previously reported in the study area [24,25]. Nonetheless, we obtained two new records for the Mexican Pacific and six new records for the CMP (Supplementary Material, Table S1), increasing the known information regarding regional crustaceans.

Several species collected during this study, i.e., *Tumidotheres margarita, Typton* sp., and *Pontonia* sp., have been reported as endosymbionts of sponges, ascidians, or bivalves. These hosts are frequently associated with pocilloporid corals, and these decapods might be recognized as having a secondary association with pocilloporid corals. *Tumidotheres margarita* is an endosymbiont of the bivalves *Barbatia reevaena, Limaria pacifica,* and *Pinctada mazatlanica* [57], which are known as *Pocillopora*-associated mollusks in the Mexican Pacific [58]. *Typton tortugae* and *T. serratus* have been recorded as being associated with sponges living on corals [59]. In this study, some sponges were found to be associated with corals, and a similar association could exist in the cases of *T. hephaestus* and *T. granulosus*. Shrimps of the genus *Pontonia* are reported as obligate symbionts of the bivalves *Pinna* spp. and *P. mazatlanica* [60]. We assumed that the *Pontonia* specimens collected during this study were dislodged from their host during the collecting process or after the samples were preserved.

Our study increased the inventory of crustaceans associated with *Pocillopora* coral in the Mexican Pacific from 59 [20–24] to 88 species. Comparatively, in Huatulco, Oaxaca, a method similar to the one used here (0.25 m^2 quadrants) recorded 47 species of brachyuran crabs in pocilloporid corals [21]. In La Paz and Loreto Bay, Baja California Sur, 44 species of decapods were recorded [22]. Furthermore, a study covering almost the entire Mexican Pacific, from the Gulf of California to Oaxaca, recorded 36 crustacean species associated with pocilloporids [24]. The difference between the number of species reported herein and by Hernández et al. [24] may be a consequence of the visual census they performed. With this method, some close species are easily confused (e.g., *Synalpheus* spp., *Trapezia* spp., and *Alpheus* spp.) or overlooked (e.g., *Hapalocarcinus marsupialis*). The expected species richness estimated by the sample-based rarefaction was 20% higher than the observed richness due to the large number of rare species collected. Expected species richness is a good indicator of the potential species expected in the area. The sample-based rarefaction confirmed that the sampling effort was sufficient to elucidate the actual number of crustacean species associated with the *Pocillopora* coral in the CMP.

Decapod crustacean fauna associated with *Pocillopora* coral has been studied in many tropical and subtropical regions of the world's oceans. The species diversity recorded in this study is superior to the 36 species associated with *Pocillopora* off the Arabian coast in the Red Sea [61]. However, it is lower than the diversity reported from Oahu (Hawaii), where 127 species were found associated with *Pocillopora damicornis* [62], and then the 91 species reported more recently in 751 colonies of *P. meandrina*, also in Oahu [3]. For the northern Great Barrier Reef, Australia, 102 species were found in 50 colonies of *P. damicornis* [63]. It is important to mention that the obligate symbiotic composition observed in our study is similar to what has been reported for the Red Sea and the Great Barrier Reef, i.e., all three studies share the same brachyuran crabs (*Trapezia bidentata, T. digitalis,* and *Domecia hispida*) and caridean shrimps (*Alpheus lottini, Synalpheus charon,* and *Harpiliopsis depressa*).

A previous study indicated that the number of species present in coral ecosystems depends on the size of the coral colony [64]. The authors reported species richness ranging from 3 to 22 per colony (1500 cm^3 size) in the Gulf of Panama; in Costa Rica, 20 cm diameter colonies had 18 species [65]. Despite using quadrants of the same size, this study collected 13–18 species and 36–711 individuals per 0.25 m^2 of coral sample. These differences in abundance and richness are substantial and cannot only be attributed to colony size.

To predict the species richness or abundance in colonies with stable conditions, some authors considered coral complexity (e.g., inter-branch space, penetration depth, and size of living space) [9], but in some cases, this factor was unable to explain the changes between different colonies [63]. For example, species such as the symbiotic *Trapezia* are not limited by coral complexity; they only need a healthy coral fragment for their survival [66]. Other characteristics, such as the percentage of live tissue and habitat degradation, could also influence the richness and abundance shifting. The species richness and abundance increase when the proportion of live coral tissue cover decreases [7,63]; this might happen because coral loss allows other species to move to new colonies. Moreover, coral mortality increases the abundance in single colonies [15], which may occur for two reasons: (1) the death of symbionts allows for other opportunistic species to move to more stable colonies, or (2) coral loss induces migrations of individuals looking for new space to live [7,9]. This situation could be happening in Punto B, where the coral colonies are isolated, fewer colonies are available, and the ecosystem is subject to anthropogenic pressure [18]. Symbiont loss does not seem to be a problem in Punto B because of the abundant obligate symbionts found in all samples.

The average taxonomic distinctness (Δ^+) varied between sites and seasons. The Δ^+ values fell inside the 95% probability funnel, meaning they were a good representation of the taxonomic diversity of decapods and stomatopods associated with pocilloporid corals. However, Chamela had a lower Δ^+ value despite having the greatest species richness among the four sites. This contrast occurred because Chamela featured the fewest supraspecific taxonomic hierarchies since many species belonged to the same families, i.e., Alpheidae (17 species) and Porcellanidae (18 species). In contrast, Punto B had the highest Δ^+ value above the global Δ^+ of the model and sustained almost the same species richness as Chamela. Punto B shared the taxonomic hierarchies with other sites and did not present any exclusive hierarchy.

Regarding the temporal variation of the taxonomic diversity, the Δ^+ values fell outside the probability funnel in the dry season, meaning a relatively low taxonomic diversity change during this season; six genera (*Areopaguristes*, *Aniculus*, *Daldorfia*, *Bottoxanthodes*, *Pontonia*, and *Pseudosquillisma*), two families (Parthenopidae and Pseudosquillidae), and one superfamily (Parthenopoidea) were not recorded in this season. In contrast, the taxonomic diversity was better represented during the wet season, when the Parthenopoidea superfamily was present, portrayed by *Daldorfia trigona*, a species not collected in the dry season. Moreover, 15 species and 6 genera were exclusively collected during the wet season (Supplementary Material, Table S1). Years one and two had similar species richness and taxonomic structure. Likewise, both Δ^+ values fell into the probability funnel close to the global Δ^+ level, demonstrating that the studied years adequately represented the taxonomic diversity estimated by the global Δ^+ model.

The nMDS ordination coupled with cluster analysis showed that Chamela had the highest taxonomic dissimilarities (Γ^+) among the studied sites. Chamela—the northernmost site—was the most different with the highest taxonomic dissimilarity, the lowest Δ^+, and the highest species richness. The Chamela samples contained one superfamily, two families, and four genera exclusive to this site, but several superfamilies, families, and genera present in the other sites were absent. It has been suggested that a low taxonomic distinctness can indicate a loss in the taxonomic diversity due to anthropogenic stress [30]. However, in this study, Punto B was the most anthropogenically affected site and displayed the highest Δ^+ values. Despite moderate disturbances, symbiotic species tend to stay in their host for a long time [9,63]. Nevertheless, some symbionts (e.g., *Trapezia*) can migrate to other coral colonies in search of more suitable habitat [66,67]. Limited habitat availability makes them pile up in the colony, increasing species richness and abundance. This phenomenon could affect the Δ^+ values in Punto B, increasing the values higher than the global Δ^+ of the model. The low levels of taxonomic diversity in Chamela might be attributed to other variables, including the spatial process [68], benthic heterogeneity, habitat availability [69], or habitat type [70]. In addition, it is important to remember that the variety of microhabitats is one of the main factors driving the diversity and abundance of coral-associated crustaceans [7].

In conclusion, the sampling effort in this study allowed for obtaining more than 70% of the expected species, indicating a good taxonomic representativity. The species richness and the taxonomic distinctness were within the expected values, despite being lower during the dry season. Most of the expected coral-obligated symbionts were collected, except for *Fennera chacei*, a small species frequently living in the coral base, which probably escaped during the collecting process. In contrast with the initial hypothesis, the sites with the most discontinuous coral cover and the largest human intervention did not have the lowest taxonomic distinctness (Punto B). However, as expected, the greatest abundance was observed in Punto B; this can be explained by the low coral availability, environmental variables, or anthropogenic stress. The present study should be complemented with α, γ, and β diversity analysis to assess the spatio–temporal differences in this particular species assemblage. It is also important to consider the influence of environmental variables, reef structural complexity, and human impact on the richness and abundance of these crustacean species, particularly in the obligate coral-symbiotic species. This study helped us to understand the crustacean assemblage associated with corals in the CMP and the spatio–temporal variations in their taxonomic diversity. Furthermore, it increased the taxonomic inventory of the coral-associated species in the studied region and the Mexican Pacific.

Supplementary Materials: The following supporting information can be downloaded at: https://www.mdpi.com/article/10.3390/d14020072/s1, Table S1: Crustacean species list organized by families, Table S2: Sample-based rarefaction results, Figure S1: Sample-based rarefaction curves for the study area, Figure S2: Individual-based rarefaction curves between sites, Figure S3: Individual-based rarefaction curves between climatic seasons and sampling years.

Author Contributions: Conceptualization and methodology, A.A.-D., F.A.R.-Z. and M.A.-P.; formal analysis and investigation, A.A.-D. and F.A.R.-Z.; resources, F.A.R.-Z.; data curation, A.A.-D. and M.A.-P.; writing–original draft preparation, A.A.-D. and F.A.R.-Z.; writing—review and editing, F.A.R.-Z., M.A.-P., E.R.-J., M.d.C.E.-G., M.E.H. and O.V.-P.; project administration and funding acquisition, F.A.R.-Z. All authors have read and agreed to the published version of the manuscript.

Funding: A.A.-D. was funded by a doctoral fellowship (371662) from the Consejo Nacional de Ciencia y Tecnología (CONACYT). The scientific research project 257987 was funded by CB2015 from CONACYT.

Institutional Review Board Statement: Not applicable.

Informed Consent Statement: Not applicable.

Data Availability Statement: The authors confirm that the data supporting the findings of this study are available within the article and its Supplementary Material. Likewise, the data are available upon request from the corresponding author.

Acknowledgments: The authors would like to thank Sharix Rubio-Bueno and Karen A. Madrigal-González for their help in the fieldwork. We thank Enrique Godínez-Domínguez (CUCSUR-U. de G.) for his support with boats during the fieldwork.

Conflicts of Interest: The authors declare no conflict of interest.

References

1. Stella, S.J.; Pratchett, M.S.; Hutchings, P.A.; Jones, G.P. Coral–associated invertebrates: Diversity, ecological importance and vulnerability to disturbance. *Oceanogr. Mar. Biol. Annu. Rev.* **2011**, *49*, 43–104.
2. Bouchet, P.; Lozouet, P.; Maestrati, P.; Heros, V. Assessing the magnitude of species richness in tropical marine environments: Exceptionally high numbers of molluscs at a New Caledonia site. *Biol. J. Linn. Soc.* **2002**, *75*, 421–436. [CrossRef]
3. Counsell, C.W.; Donahue, M.J.; Edwards, K.F.; Franklin, E.C.; Hixon, M.A. Variation in coral-associated cryptofaunal communities across spatial scales and environmental gradients. *Coral Reefs* **2018**, *37*, 827–840. [CrossRef]
4. Thiel, M.; Baeza, J.A. Factors affecting the social behaviour of crustaceans living symbiotically with other marine invertebrates: A modelling approach. *Symbiosis* **2001**, *30*, 163–190.

5. Castro, P. Symbiotic Brachyura. In *Treatise on Zoology-Anatomy, Taxonomy, Biology. The Crustacea Volume 9*; Castro, P., Davie, P., Guinot, D., Schram, F.R., von Vaupel Klein, J.C., Eds.; Brill: Leiden, The Netherlands, 2015; pp. 543–581. [CrossRef]

6. Abele, L.G. Comparative species composition and relative abundance of decapod crustaceans in marine habitats of Panama. *Mar. Biol.* **1976**, *38*, 263–278. [CrossRef]

7. González-Gómez, R.; Briones-Fourzán, P.; Álvarez-Filip, L.; Lozano-Álvarez, E. Diversity and abundance of conspicuous macrocrustaceans on coral reefs differing in level of degradation. *PeerJ* **2018**, *6*, e4922. [CrossRef] [PubMed]

8. Stewart, H.L.; Holbrook, S.J.; Schmitt, R.J.; Brooks, A.J. Symbiotic crabs maintain coral health by clearing sediments. *Coral Reefs* **2006**, *25*, 609–615. [CrossRef]

9. Head, C.E.; Bonsall, M.B.; Koldewey, H.; Pratchett, M.S.; Speight, M.; Rogers, A.D. High prevalence of obligate coral-dwelling decapods on dead corals in the Chagos Archipelago, central Indian Ocean. *Coral Reefs* **2015**, *34*, 905–915. [CrossRef]

10. Glynn, P.W. Defense by symbiotic crustacea of host corals elicited by chemical cues from predator. *Oecologia* **1980**, *47*, 287–290. [CrossRef] [PubMed]

11. McKeon, C.S.; Moore, J.M. Species and size diversity in protective services offered by coral guard-crabs. *PeerJ* **2014**, *2*, e574. [CrossRef] [PubMed]

12. McKeon, C.S.; Stier, A.C.; McIlroy, S.E.; Bolker, B.M. Multiple defender effects: Synergistic coral defense by mutualist crustaceans. *Oecologia* **2012**, *169*, 1095–1103. [CrossRef]

13. Britayev, T.A.; Mikheev, V. Clumped spatial distribution of scleractinian corals influences the structure of their symbiotic associations. *Biol. Sci.* **2013**, *448*, 45–48. [CrossRef]

14. Randall, J.E. Food habits of reef fishes of the West Indies. *Stud. Trap. Oceanogr.* **1967**, *5*, 665–847.

15. Enochs, I.C. Motile cryptofauna associated with live and dead coral substrates: Implications for coral mortality and framework erosion. *Mar. Biol.* **2012**, *159*, 709–722. [CrossRef]

16. Leray, M.; Meyer, C.P.; Mills, S.C. Metabarcoding dietary analysis of coral dwelling predatory fish demonstrates the minor contribution of coral mutualists to their highly partitioned, generalist diet. *PeerJ* **2015**, *3*, e1047. [CrossRef]

17. Reyes-Bonilla, H. Coral reefs of the Pacific coast of Mexico. In *Latin American Coral Reefs*; Cortés, J., Ed.; Elsevier: Amsterdam, The Netherlands, 2003; pp. 331–349. [CrossRef]

18. Hernández-Zulueta, J.; Rodríguez-Zaragoza, F.A.; Araya, R.; Vargas-Ponce, O.; Rodríguez-Troncoso, A.P.; Cupul-Magaña, A.L.; Díaz-Pérez, L.; Ríos-Jara, E.; Ortiz, M. Multi-scale analysis of hermatypic coral assemblages at Mexican Central Pacific. *Sci. Mar.* **2017**, *81*, 91–102. [CrossRef]

19. Pereyra Ortega, R.T. Cangrejos Anomuros y Braquiuros (Crustacea: Decapoda) Simbiontes del Coral *Pocillopora elegans* de Los Islotes, Baja California Sur, Mexico. Undergraduate Thesis, Universidad Autónoma de Baja California Sur, La Paz, Mexico, 1998.

20. Hernández, L. Camarones Stenopodideos y Carideos (Crustacea: Decapoda: Stenopodidae, Caridea) Asociados a Colonias de Coral, en Los Islotes, B.C.S., México. Undergraduate Thesis, Universidad Autónoma de Baja California Sur, La Paz, Mexico, 1999.

21. Ramírez-Luna, S.; De La Cruz-Agüero, G.; Barrientos-Luján, N.A. Variación espacio-temporal de Porcellanidae, Majoidea y Xanthoidea asociados a corales del género *Pocillopora* en Bahías de Huatulco, México. In *Contribuciones al Estudio de Los Crustáceos del Pacífico Este [Contributions to the Study of East Pacific Crustaceans]*; Hendrickx, M.E., Ed.; Instituto de Ciencias del Mar y Limnología: Ciudad de México, México, 2002; pp. 233–254.

22. Hernández, L.; Balart, E.F.; Reyes-Bonilla, H. Effect of hurricane John (2006) on the invertebrates associated with corals in Bahia de la Paz, Gulf of California. In Proceedings of the 11th International Coral Reef Symposium, Ft. Lauderdale, FL, USA, 7–11 July 2008.

23. Hernández, L.; Reyes-Bonilla, H.; Balart, E.F. Efecto del blanqueamiento del coral por baja temperatura en los crustáceos decápodos asociados a arrecifes del suroeste del Golfo de California. *Rev. Mex. Biodivers.* **2010**, *81*, 113–119. [CrossRef]

24. Hernández, L.; Ramírez-Ortiz, G.; Reyes-Bonilla, H. Coral–associated decapods (Crustacea) from the Mexican Tropical Pacific coast. *Zootaxa* **2013**, *3609*, 451–464. [CrossRef]

25. Ayón-Parente, M.; Hermoso-Salazar, M.; Hendrickx, M.E.; Galván-Villa, C.M.; Ríos-Jara, E.; Bastida-Izaguirre, D. The caridean shrimps (Crustacea: Decapoda: Caridea: Alpheoidea, Palaemonoidea, and Processoidea) from Bahía Chamela, Mexico. *Rev. Mex. Bio.* **2016**, *87*, 311–327. [CrossRef]

26. Clarke, K.R.; Warwick, R.M. The taxonomic distinctness measure of biodiversity: Weighting of step lengths between hierarchical levels. *Mar. Ecol. Prog. Ser.* **1999**, *184*, 21–29. [CrossRef]

27. Magurran, E.A. *Measuring Biological Diversity*; Blackwell Publishing: Oxford, UK, 2004; p. 248.

28. Clarke, K.R.; Warwick, R.M. A further biodiversity index applicable to species lists: Variation in taxonomic distinctness. *Mar. Ecol. Prog. Ser.* **2001**, *216*, 265–278. [CrossRef]

29. Ji, L.; Jiang, X.; Liu, C.; Xu, Z.; Wang, J.; Qian, S.; Zhou, H. Response of traditional and taxonomic distinctness diversity indices of benthic macroinvertebrates to environmental degradation gradient in a large Chinese shallow lake. *Environ. Sci. Pollut. Res.* **2020**, *27*, 21804–21815. [CrossRef]

30. Leonard, D.R.P.; Clarke, K.R.; Somerfield, P.J.; Warwick, R.M. The application of an indicator based on taxonomic distinctness for UK marine biodiversity assessment. *J. Environ. Manag.* **2006**, *78*, 52–62. [CrossRef]

31. Roger, S.I.; Clarke, K.R.; Reynolds, J.D. The taxonomic distinctness of coastal bottom-dwelling fish communities of the North-east Atlantic. *J. Anim. Ecol.* **1999**, *68*, 769–792. [CrossRef]

32. Jiang, X.; Pan, B.; Sun, Z.; Cao, L.; Lu, Y. Application of taxonomic distinctness indices of fish assemblages for assessing effects of river-lake disconnection and eutrophication in floodplain lakes. *Ecol. Indic.* **2020**, *110*, 105955. [CrossRef]

33. Heino, J.; Soininen, J.; Lappalainen, J.; Virtanen, R. The relationship between species richness and taxonomic distinctness in freshwater organisms. *Limnol. Oceanogr.* **2005**, *50*, 978–986. [CrossRef]

34. Jiang, X.; Song, Z.; Xiong, J.; Xie, Z. Can excluding non-insect taxa from stream macroinvertebrate surveys enhance the sensitivity of taxonomic distinctness indices to human disturbance? *Ecol. Indic.* **2014**, *41*, 175–182. [CrossRef]

35. Heino, J.; Mykrä, H.; Hämäläinen, H.; Aroviita, J.; Muotka, T. Responses of taxonomic distinctness and species diversity indices to anthropogenic impacts and natural environmental gradients in stream macroinvertebrates. *Freshw. Biol.* **2007**, *52*, 1846–1861. [CrossRef]

36. Heino, J.; Alahuhta, J.; Fattorini, S. Phylogenetic diversity of regional beetle faunas at high latitudes: Patterns, drivers and chance along ecological gradients. *Biodivers. Conserv.* **2015**, *24*, 2751–2767. [CrossRef]

37. López-Pérez, A.; Granja-Fernández, R.; Benítez-Villalobos, F.; Jiménez-Antonio, O. *Pocillopora damicornis*-associated echinoderm fauna: Richness and community structure across the southern Mexican Pacific. *Marine Biodiversity* **2017**, *47*, 481–490. [CrossRef]

38. Spalding, M.D.; Fox, H.E.; Allen, G.R.; Davidson, N.; Ferdaña, Z.A.; Finlayson, M.; Halpern, B.S.; Jorge, M.A.; Lombana, A.; Lourie, S.A.; et al. Marine Ecoregions of the world: A bioregionalization of coastal and shelf areas. *BioScience* **2007**, *57*, 573–583. [CrossRef]

39. Kessler, W.S. The circulation of the eastern tropical Pacific: A review. *Prog. Oceanogr.* **2006**, *69*, 181–217. [CrossRef]

40. Fiedler, P.C.; Talley, L.D. Hydrography of the eastern tropical Pacific: A review. *Prog. Oceanogr.* **2006**, *69*, 143–180. [CrossRef]

41. Fiedler, P.C.; Lavín, M.F. Oceanographic conditions of the eastern tropical Pacific. In *Coral Reefs of the Eastern Tropical Pacific*, 1st ed.; Glynn, P.W., Manzello, D.O., Enochs, I.C., Eds.; Springer: Dordrecht, The Netherlands, 2017; Volume 1, pp. 59–83.

42. Rathbun, M.J. The cancroid crabs of America of the families Euryalidae, Portunidae, Atelecyclidae, Cancridae, and Xanthidae. *Bull. U. S. Natl. Mus.* **1930**, *152*, 1–609. [CrossRef]

43. Haig, J. The Porcellanidae (Crustacea Anomura) of the eastern Pacific. In *Allan Hancock Pacific Expedition*; University of Southern California: Los Angeles, CA, USA, 1960; Volume 24, pp. 1–440.

44. Abele, L.G.; Kim, W. *An Illustrated Guide to the Marine Decapod Crustaceans of Florida Volume 8*; Technical Series; Department of Environmental Regulations, Florida State University: Tallahassee, FL, USA, 1986.

45. Castro, P. Eastern Pacific species of *Trapezia* (Crustacea, Brachyura: Trapeziidae), sibling species symbiotic with reef corals. *Bull. Mar. Sci.* **1996**, *58*, 531–554.

46. Anker, A.; Ahyong, S.T.; Noël, P.Y.; Palmer, A.R. Morphological phylogeny of alpheid shrimps: Parallel preadaptation and the origin of a key morphological innovation, the snapping claw. *Evolution* **2006**, *60*, 2507–2528. [CrossRef] [PubMed]

47. Hendrickx, M.E.; Ayón-Parente, M.; Félix-Pico, E.; Vargas-López, G. Hermit crabs (Crustacea: Paguroidea) in the biological collection of CICIMAR, Instituto Politécnico Nacional, La Paz, Baja California Sur, México. In *Contribution to the Study of East Pacific Crustaceans*; Hendrickx, M.E., Ed.; Instituto de Ciencias del Mar y Limnología, Universidad Nacional Autónoma de México: Ciudad de México, México, 2008; Volume 5, pp. 17–21.

48. Ayón-Parente, M. Taxonomía, Zoogeografía y Aspectos Ecológicos de los Cangrejos Ermitaños de la Familia Diogenidae (Crustacea: Decapoda: Anomura) del Pacífico Mexicano. Ph.D. Thesis, Posgrado en Ciencias del Mar y Limnología, Instituto de Ciencias del Mar y Limnología, Universidad Nacional Autónoma de México, Ciudad de México, México, 2009.

49. García-Madrigal, M.S.; Andréu-Sánchez, L.I. Los cangrejos porcelánidos (Decapoda: Anomura) del Pacífico sur de México, incluyendo una lista y clave de identificación para todas las especies del Pacífico oriental tropical. *Ciencia y Mar* **2009**, *13*, 23–54.

50. Hermoso-Salazar, A.M. Sistemática y Biogeografía del género *Synalpheus* (Decapoda: Caridea) del Pacífico Oriental. Ph.D. Thesis, Universidad Nacional Autónoma de México, Mexico City, México, 2009.

51. Salgado-Barragán, J.; Hendrickx, M.E. Clave ilustrada para la identificación de los estomatópodos (Crustacea: Hoplocarida) del Pacífico oriental. *Rev. Mex. Biodivers.* **2010**, *81*, 1–49. [CrossRef]

52. Hiller, A.; Lessios, H.A. Marine species formation along the rise of Central America: The anomuran crab *Megalobrachium*. *Mol. Ecol.* **2020**, *29*, 413–428. [CrossRef]

53. González-Oreja, J.; De la Fuente-Díaz-Ordaz, A.A.; Hernández-Santín, L.; Buzo-Franco, D.; Bonache-Regidor, C. Evaluación de estimadores no paramétricos de la riqueza de especies. Un ejemplo con aves en áreas verdes de la ciudad de Puebla, México. *Anim. Biodivers. Conserv.* **2010**, *33*, 31–45.

54. Collwell, R. Estimate S: Statistical Estimation of Species Richness and Shared Species from Samples. 2015. Available online: viceroy.colorado.edu/estimates/EstimateSPages/EstimateSRegistration.htm (accessed on 9 January 2020).

55. Warwick, R.M.; Clarke, K.R. New biodiversity measures reveal a decrease in taxonomic distinctness with increasing stress. *Mar. Ecol. Prog. Ser.* **1995**, *129*, 301–305. [CrossRef]

56. Clarke, K.R.; Gorley, R.N. *Primer v7: User Manual/Tutorial*; PRIMER-E: Plymouth, UK, 2006.

57. de Gier, W.; Becker, C. A Review of the Ecomorphology of Pinnotherine Pea Crabs (Brachyura: Pinnotheridae), with an Updated List of Symbiont-Host Associations. *Diversity* **2020**, *12*, 431. [CrossRef]

58. Barrientos-Lujan, N.A.; Rodríguez-Zaragoza, F.A.; López-Pérez, A. Richness, abundance and spatial heterogeneity of gastropods and bivalves in coral ecosystems across the Mexican Tropical Pacific. *J. Molluscan Stud.* **2021**, *87*, eyab004. [CrossRef]

59. Holthuis, L.B. *A General Revision of the Palaemonidae (Crustacea Decapoda Natantia) of the Americas*; University of Southern California: Los Angeles, CA, USA, 1951; Volume 11, pp. 1–332.

60. Fransen, C.H. Taxonomy, phylogeny, historical biogeography, and historical ecology of the genus *Pontonia* (Crustacea: Decapoda: Caridea: Palaemonidae). *Zool. Verh.* **2002**, *336*, 1–433.

61. Britayev, T.A.; Spiridonov, V.A.; Deart, Y.V.; El-Sherbiny, M. Biodiversity of the community associated with *Pocillopora verrucosa* (Scleractinia: Pocilloporidae) in the Red Sea. *Mar. Bio.* **2017**, *47*, 1093–1109. [CrossRef]

62. Coles, S. Species diversity of decapods associated with living and dead reef coral *Pocillopora meandrina*. *Mar. Ecol. Prog. Ser.* **1980**, *2*, 281–291. [CrossRef]

63. Stella, J.S.; Jones, G.P.; Pratchett, M.S. Variation in the structure of epifaunal invertebrate assemblages among coral hosts. *Coral Reefs* **2010**, *29*, 957–973. [CrossRef]

64. Abele, L.G.W.; Patton, K. The size of coral heads and the community biology of associated decapod crustaceans. *J. Biogeogra.* **1976**, *3*, 35–47. [CrossRef]

65. Alvarado, J.J.; Vargas-Castillo, R. Invertebrados asociados al coral constructor de arrecifes *Pocillopora damicornis* en Playa Blanca, Bahía Culebra, Costa Rica. *Rev. Biol. Trop.* **2012**, *60*, 77–92. [CrossRef]

66. Canizales-Flores, H.M.; Rodríguez-Troncoso, A.P.; Rodríguez-Zaragoza, F.A.; Cupul-Magaña, A.L. A Long-term symbiotic relationship: Recruitment and fidelity of the crab *Trapezia* on its coral host *Pocillopora*. *Diversity* **2021**, *13*, 450. [CrossRef]

67. Castro, P. Movements between coral colonies in *Trapezia ferruginea* (Crustacea: Brachyura), an obligate symbiont of scleractinian corals. *Mar. Biol.* **1978**, *46*, 237–245. [CrossRef]

68. Bhat, A.; Magurran, A.E. Taxonomic distinctness in a linear system: A test using a tropical freshwater fish assemblage. *Ecography* **2006**, *29*, 104–110. [CrossRef]

69. Leira, M.; Chen, G.; Dalton, C.; Irvine, K.; Taylor, D. Patterns in freshwater diatom taxonomic distinctness along an eutrophication gradient. *Freshw. Biol.* **2009**, *54*, 1–14. [CrossRef]

70. Bevilacqua, S.; Fraschetti, S.; Musco, L.; Guarnieri, G.; Terlizzi, A. Low sensi-tiveness of taxonomic distinctness indices to human impacts: Evidences acrossmarine benthic organisms and habitat types. *Ecol. Indic.* **2011**, *11*, 448–455. [CrossRef]

Interesting Images

Black Mantle Tissue of Endolithic Mussels (*Leiosolenus* spp.) Is Cloaking Borehole Orifices in Caribbean Reef Corals

Bert W. Hoeksema [1,2,3,*], Annabel Smith-Moorhouse [1,2], Charlotte E. Harper [1,2], Roel. J. van der Schoot [1,2], Rosalie F. Timmerman [1,2], Roselle Spaargaren [1,2] and Sean J. Langdon-Down [1,2]

[1] Taxonomy, Systematics and Geodiversity Group, Naturalis Biodiversity Center, P.O. Box 9517, 2300 RA Leiden, The Netherlands; a.smith-moorhouse@student.rug.nl (A.S.-M.); c.harper@student.rug.nl (C.E.H.); roel.vanderschoot@naturalis.nl (R.J.v.d.S.); r.f.timmerman@student.rug.nl (R.F.T.); r.spaargaren@student.rug.nl (R.S.); s.langdon-down@student.rug.nl (S.J.L.-D.)

[2] Groningen Institute for Evolutionary Life Sciences, University of Groningen, P.O. Box 11103, 9700 CC Groningen, The Netherlands

[3] Institute of Biology Leiden, Leiden University, P.O. Box 9505, 2300 RA Leiden, The Netherlands

* Correspondence: bert.hoeksema@naturalis.nl

Abstract: Bioerosion caused by boring mussels (Mytilidae: Lithophaginae) can negatively impact coral reef health. During biodiversity surveys of coral-associated fauna in Curaçao (southern Caribbean), morphological variation in mussel boreholes was studied. Borings were found in 22 coral species, 12 of which represented new host records. Dead corals usually showed twin siphon openings, for each mussel shaped like a figure of eight, which were lined with a calcareous sheath and protruded as tubes from the substrate surface. Most openings surrounded by live coral tissue were deeper and funnel-shaped, with outlines resembling dumbbells, keyholes, ovals or irregular ink blotches. The boreholes appeared to contain black siphon and mantle tissue of the mussel. Because of the black color and the hidden borehole opening in live host corals, the mantle tissue appeared to mimic dark, empty holes, while they were actually cloaking live coral tissue around the hole, which is a new discovery. By illustrating the morphological range of borehole orifices, we aim to facilitate the easy detection of boring mussels for future research.

Keywords: bioerosion; boring; coral health; Curaçao; host records; Lithophaginae; Mytilidae

Citation: Hoeksema, B.W.; Smith-Moorhouse, A.; Harper, C.E.; van der Schoot, R.J.; Timmerman, R.F.; Spaargaren, R.; Langdon-Down, S.J. Black Mantle Tissue of Endolithic Mussels (*Leiosolenus* spp.) Is Cloaking Borehole Orifices in Caribbean Reef Corals. *Diversity* **2022**, *14*, 401. https://doi.org/10.3390/d14050401

Academic Editor: Marco Taviani

Received: 5 May 2022
Accepted: 18 May 2022
Published: 20 May 2022

Publisher's Note: MDPI stays neutral with regard to jurisdictional claims in published maps and institutional affiliations.

Boring mussels (Mytilidae: Lithophaginae) are notorious for their bioerosion of limestone rock, bivalve shells, reef corals and various manmade calcareous substrates [1–9]. Most of these boring mussels (also called date mussels) belong to the genera *Leiosolenus* Carpenter, 1857 and *Lithophaga* Röding, 1798 [10]. In addition to causing damage to the structure of reef corals [11–15], these animals are suspected to make host corals more susceptible to diseases [16].

In order to detect the presence of boring mussels inside corals, it is important to recognize the orifices of their boreholes. For their feeding and respiration, boring mussels inhale and exhale seawater through a pair of siphons at the posterior edge of their mantle tissue [17–19]. The siphons use openings in the substrate surface for contact with the surrounding seawater [20]. The outline of such openings is described as "figure-of-eight" or "dumbbell" [3–5,8,20–23] shape, not to be confused with the twin openings of U-shaped excavations of *Polydora* worms (Polychaeta) [24–28] and the perforations made by boring clionaid sponges (Porifera) [25,29].

In mussels of the genus *Leiosolenus*, the borehole and its openings are lined with an aragonite (calcareous) sheath that is excreted by the bivalve [16,20,30,31]. At the substrate surface, such sheaths may appear as chimney-like tubes that provide protection to the siphons [21,22,32,33]. However, these sheaths are not always visible, and the openings of some borings are described as being oval in shape, which may perhaps be influenced by the host coral or by overgrowing algae [20,32,33]. Oval orifices of mussel borings can be irregular in shape [32,34] and should not be confused with the crescent-shaped openings of some coral-dwelling gall-crab species [35–37]. Owing to their morphological variability, the openings of mussel holes may not always be recognized; it is possible that they therefore become classified as "unknown holes" [38]. Because boring mussels can have a negative impact on the health of reef corals [12,16], it is important that their presence can be detected through the easy recognition of their orifices. In this study, we provide information on how these openings can be spotted in the field.

During biodiversity surveys of coral-associated fauna along the leeward side of Curaçao (southern Caribbean) in October–December 2021 and April 2022 [39,40] a number of live and dead corals were checked for boreholes of lithophagine mussels. To verify the presence of mussels underneath openings, two corals were broken to reveal the position of the mussels (Figure 1).

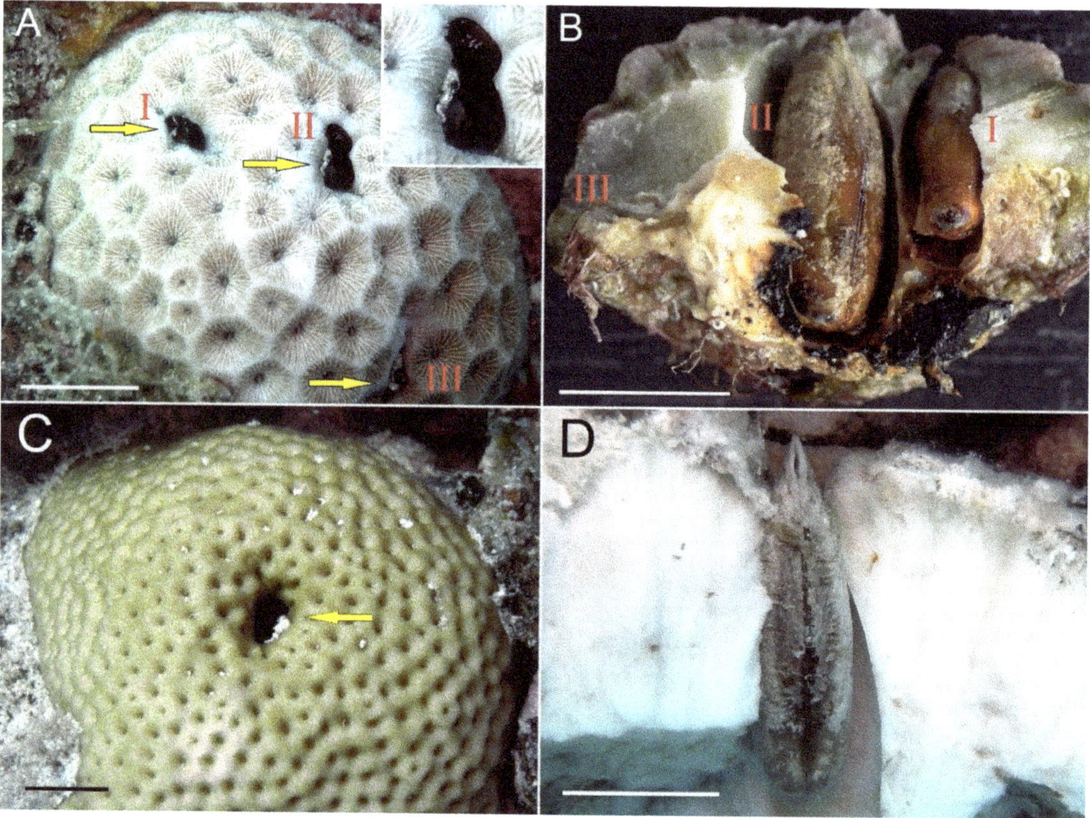

Figure 1. Coral colonies of *Siderastrea siderea* at Curaçao, showing the position of borehole openings (**A,C**: arrows) and *Leiosolenus* mussels underneath them (**B,D**). One coral contains three mussels (**A,B**: I–III) and the other only one (**C,D**). Each exposed mussel has the posterior side upward, showing either a lateral side (**B**) or the dorsal side (**D**). The dark color of each hole (**A,C**) indicates the presence of the mussel's mantle tissue; in some individuals approaching dark Bordeaux red (**A**: insert 2× enlargement). The mantle tissue may be covered by some detritus particles (**B,D**). In exposed mussels, the mantle tissue is retracted inside the shell (**B,D**). Scale bars: 1 cm.

The morphological variety in the orifices appeared to be more extensive than previously reported. Many corals, mostly dead but also live ones, showed two calcareous tubes (sheaths), protruding from the substrate surface, described as aragonite chimney-like structures [21,32]. In addition to showing a figure-of-eight shape consisting of two connected tubes (Figure 2A–E), some twin openings appeared to be separate (Figure 2F). A slit was seen in the calcareous margin where twin tubes were merged, varying in width (Figure 2A–E). The tubes did not protrude as high as those made by boring bivalves of the family Gastrochaenidae, which excavate in dead coral [22,32]. Most orifices in live corals showed a so-called "dumbbell shape", although "keyhole shape" appears to be more appropriate (Figure 2G–I). Other openings surrounded by live coral tissue appeared to have an oval outline (Figure 2J–L) or one resembling an irregular ink blotch (Figure 2M,N). A few boreholes showed an empty *Leiosolenus* shell inside (Figure 2O).

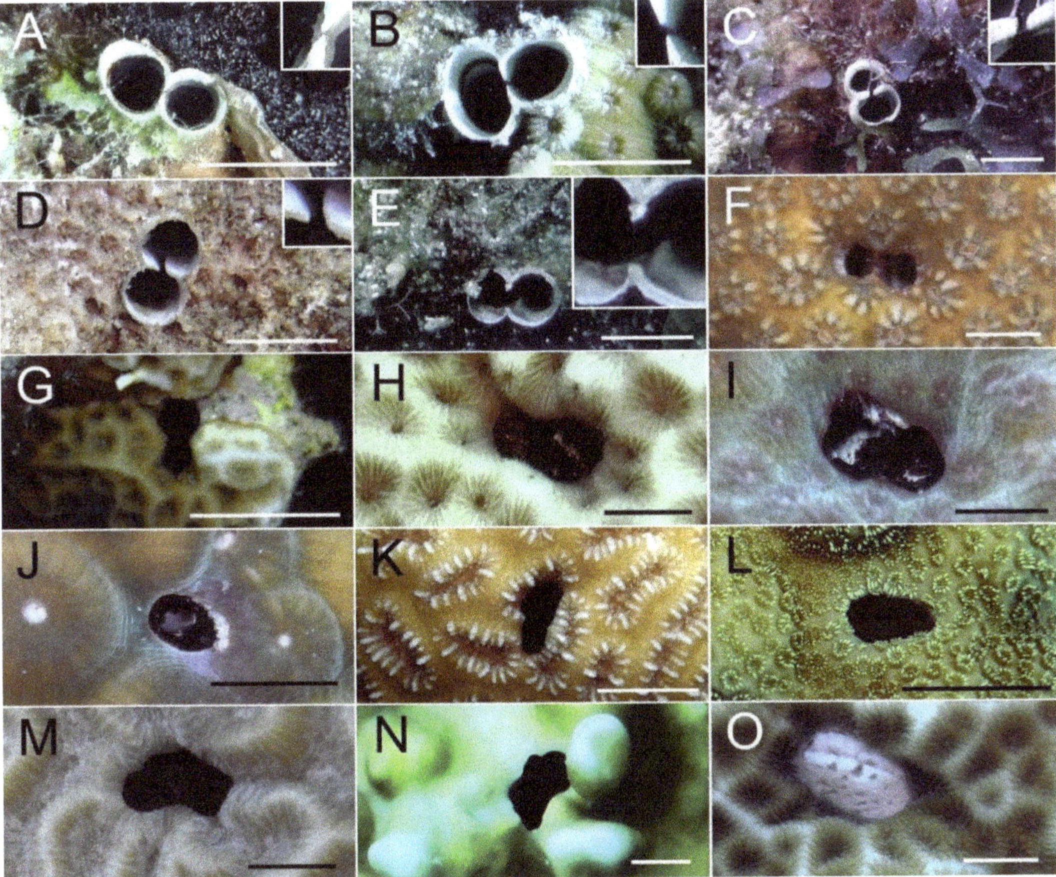

Figure 2. Morphological variation of orifices in corals containing *Leiosolenus* mussels at Curaçao. (**A–E**) Figure-of-eight shape with two calcareous tubes showing black siphon tissue inside; the tubes are connected apart from a slit (inserts: 2.5× enlargement). (**F**) The siphon tubes are separated by the host coral. Black mantle tissue is cloaking holes that are shaped like a dumbbell or keyhole (**G–I**), an oval (**J–L**), or an irregular ink blotch (**M,N**). (**O**) A hole containing valves of a dead mussel. Substrate: dead coral (**A,C,D**); live corals of *Orbicella franksi* (**B,I**), *Montastraea cavernosa* (**E,J**), *Madracis senaria* (**F**), *Agaricia humilis* (**G,O**), *Siderastrea siderea* (**H**), *Favia fragum* (**K**), *Porites astreoides* (**L**), *Pseudodiploria strigosa* (**M**), *Millepora alcicornis* (**N**). Scale bars: 1 cm.

The inner surface of the tubes was lined with black siphon tissue (Figure 2A–F). The tubes were not visible in the larger holes (Figure 2G–N), which appeared to be pitch black, making them appear to be empty. Closer inspection showed that they were filled with the mussel's black mantle and siphon tissue. Disturbance evoked the retraction of the tissue, revealing that the orifice was funnel-shaped (Figure 3) and that the mantle originally covered live polyps around the hole, masking its true outline. Since the boring activity of the mussels is in posterior and lateral directions [22,41] and the host coral expands, the mussels are forced to move their holes upward in order to remain close to the host's surface [22], as illustrated by Gohar and Soliman (1983: Figure 11B) [23] and by Yahel et al. (2009: Figure 1B) [42]. It is notable that boring mussels of some genera have anterior boring glands [20], suggesting that they can indeed bore in an upward direction. When the calcareous tubes fail in keeping track of the expanding coral and stop reaching the host's surface, the mussel's mantle sustains an open orifice surrounded by growing coral tissue, forcing the host to form a funnel-shaped entrance (Figures 1C, 2H,I, and 3). Such openings may resemble crevices formed by *Pedum* clams that live inside massive corals [11,43,44] or incavations formed by some coral-gall crabs [35].

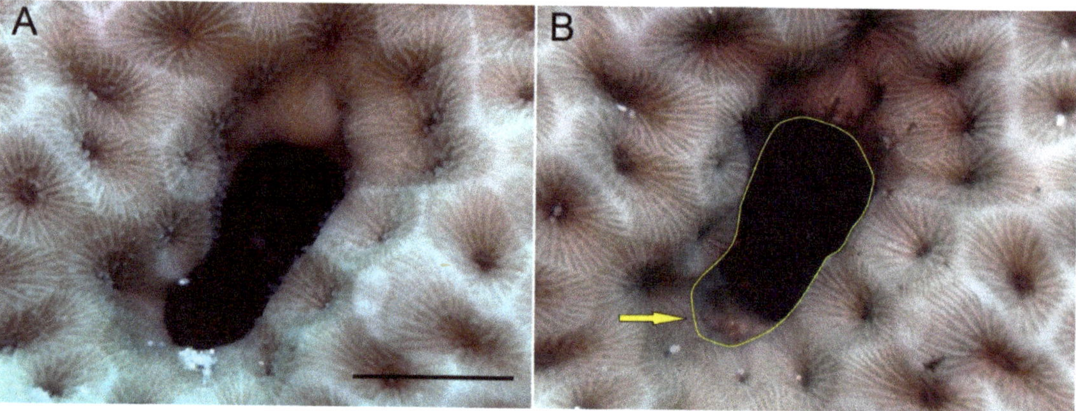

Figure 3. *Leiosolenus* boring in a colony of *Siderastrea siderea* in Curaçao with a funnel-shaped entrance. (**A**) The mussel's black mantle tissue expanded with a keyhole-shaped outline. (**B**) The same borehole (from a slightly different angle) with part of the mantle tissue withdrawn (arrow and contour line showing the previous position as depicted in (**A**)). Retraction of the mantle tissue reveals even more that the opening is funnel shaped. Scale bar: 0.5 cm.

Close up, the color of the mantle tissue appeared to be dark red (Bordeaux) in some mussel individuals, which is slightly visible in Figure 1A. An examination of black holes in corals for the presence of mantle tissue inside makes it easier to see whether boring mussels are present, distinguishing them from dark empty holes without mussels. Previous studies on boring mussels did not pay attention to how mantle coloration may cause lithophagine holes to become less discernible. This finding may help to study whether coral-dwelling date mussels are more abundant than previously thought.

Mussel boreholes were found in 20 scleractinian species and two milleporids (Table 1). Twenty species had large holes (oval and other shapes), and only nine showed figure-of-eight twin openings (Table 1). There were twelve new Caribbean host records, including those of the two *Millepora* species. Three extant *Leiosolenus* species have been described from Caribbean corals [45–49]: *L. aristatus* (Dillwyn, 1817), *L. bisulcatus* (d'Orbigny, 1853) and *L. dixonae* (Scott, 1986). The latter has only been recorded from three *Madracis* species: *M. auretenra* (misidentified as *M. mirabilis*), *M. decactis* (Lyman, 1859) and *M. formosa* Wells, 1973 [45].

Table 1. Coral species at Curaçao observed as hosts for *Leiosolenus*; * = new host record. Shape of orifices observed: T = figure of eight; O = other (oval, dumbbell, keyhole and ink blotch).

Host Taxon	Orifice Shape
Cnidaria: Anthozoa: Scleractinia	
Agariciidae	
Agaricia agaricites (Linnaeus, 1758)	O
Agaricia humilis (Verrill, 1901) *	T O
Agaricia lamarcki Milne Edwards & Haime, 1851 *	O
Astrocoeniidae	
Stephanocoenia intersepta (Esper, 1795)	O
Faviidae: Faviinae	
Colpophyllia natans (Houttuyn, 1772) *	T O
Diploria labyrinthiformis (Linnaeus, 1758) *	
Favia fragum (Esper, 1793)	O
Pseudodiploria strigosa (Dana, 1846)	T O
Meandrinidae	
Eusmilia fastigiata (Pallas, 1766) *	O
Meandrina meandrites (Linnaeus, 1758) *	O
Merulinidae	
Orbicella annularis (Ellis & Solander, 1786)	O
Orbicella faveolata (Ellis & Solander, 1786) *	T O
Orbicella franksi (Gregory, 1895) *	T O
Montastraeidae	
Montastraea cavernosa (Linnaeus, 1767)	O
Pocilloporidae	
Madracis auretenra Locke, Weil & Coates, 2007	O
Madracis decactis (Lyman, 1859)	T O
Madracis pharensis (Heller, 1868) *	T
Madracis senaria Wells, 1973 *	T O
Poritidae	
Porites astreoides Lamarck, 1816	O
Siderastreidae	
Siderastrea siderea (Ellis & Solander, 1768)	O
Cnidaria: Hydrozoa: Anthoathecata	
Milleporidae	
Millepora alcicornis Linnaeus, 1758 *	O
Millepora complanata Lamarck, 1816 *	T O
Dead coral	T

Leiosolenus aristatus has been recorded from Brazil as an introduced species in invasive *Tubastraea* corals [50] and also from Southeast Florida but without a host record [51]. *Leiosolenus bisulcatus* was previously recorded from *Agaricia agaricites*, *Favia fragum*, *Pseudodiploria strigosa*, *Siderastrea radians*, *Siderastrea siderea* and *Stephanocoenia intersepta* (as *S. michelini*) [45,47,52]. *Leiosolenus bisulcatus* has also been recorded from *Oculina arbuscula* Agassiz, 1880 in North Carolina, USA [53] and from *Mussismilia hispida* (Verrill, 1902) and *Siderastrea stellata* Verrill, 1868 in Brazil [54]. In the present study, the mussels were not identified at the species level, but considering previous host records, *L. bisulcatus* is the most likely an associate for most host coral species, with the exception of *L. aristatus* for *Madracis*.

By presenting the host range of boring mussels and by showing the morphological range of their borehole orifices, we aim to facilitate the easy detection of these bioeroding organisms in future research. Our findings may also help in the interpretation of fossil holes of boring mussels, recognized as trace fossils of the ichnogenus *Gastrochaenolites*, and may tell us more about the condition and habitat of their host corals or other substrates when these were still alive [4,55–59].

For a better understanding of the host specificity of coral-associated boring mussels, more research is needed on the host selection during settlement of their larvae, like in earlier studies on Indo-Pacific Lithophaginae [31,60,61], some coral barnacles [62,63], and

Christmas tree worms [64,65]. The present findings may stimulate future studies on borehole orifices in the Indo-Pacific, where more species of coral-dwelling Lithophaginae and host-coral species occur than in the Atlantic [32,66–75]. Molecular techniques are available [2,18,73,76] to study the host specificity of coral-dwelling Lithophaginae on coral reefs in both the Atlantic and the Indo-Pacific.

Coral-dwelling mussels are not the only invertebrates participating in the coral-associated biodiversity of reef corals [77–81]. It is noteworthy that Lithophaginae may also contribute to this fauna indirectly by acting as hosts for symbiotic species themselves, such as pea crabs [82,83]. It is evident that more research is needed on the ecology and evolution of coral-dwelling mussels.

Author Contributions: Conceptualization, B.W.H. and R.J.v.d.S.; methodology, B.W.H., C.E.H., S.J.L.-D., R.J.v.d.S., A.S.-M., R.S. and R.F.T.; validation, B.W.H.; formal analysis, B.W.H.; investigation, B.W.H., C.E.H., S.J.L.-D., R.J.v.d.S., A.S.-M., R.S. and R.F.T.; resources, B.W.H., C.E.H., S.J.L.-D., R.J.v.d.S., A.S.-M., R.S. and R.F.T.; data curation, B.W.H., C.E.H., S.J.L.-D., R.J.v.d.S., A.S.-M., R.S. and R.F.T.; writing—original draft preparation, B.W.H.; writing—review and editing, B.W.H., C.E.H., S.J.L.-D., R.J.v.d.S., A.S.-M., R.S. and R.F.T.; visualization, B.W.H., C.E.H., S.J.L.-D., R.J.v.d.S., A.S.-M., R.S. and R.F.T.; supervision, B.W.H.; project administration, B.W.H.; funding acquisition, B.W.H., C.E.H., S.J.L.-D., R.J.v.d.S., A.S.-M., R.S. and R.F.T. All authors have read and agreed to the published version of the manuscript.

Funding: The field research in Curaçao was funded by the Alida M. Buitendijk Fund, the Jan-Joost ter Pelkwijk Fund, the Holthuis Fund, the Groningen University Fund, the Treub Maatschappij (Society for the Advancement of Research in the Tropics) and the Dutch Research Council (NWO) Doctoral Grant for Teachers Programme (nr. 023.015.036).

Institutional Review Board Statement: Not applicable.

Informed Consent Statement: Not applicable.

Data Availability Statement: Not applicable.

Acknowledgments: We are grateful to the funding agencies mentioned above. We thank the staff of CARMABI (Curaçao) and the Dive Shop for their hospitality and assistance during the fieldwork. We want to thank four anonymous reviewers for their constructive comments, which helped us to improve the manuscript.

Conflicts of Interest: The authors declare no conflict of interest.

References

1. Jones, B.; Pemberton, S.G. *Lithophaga* borings and their influence on the diagenesis of corals in the Pleistocene Ironshore Formation of Grand Cayman Island, British West Indies. *Palaios* **1988**, *3*, 3–21. [CrossRef]
2. Owada, M. Functional morphology and phylogeny of the rock-boring bivalves *Leiosolenus* and *Lithophaga* (Bivalvia: Mytilidae): A third functional clade. *Mar. Biol.* **2007**, *150*, 853–860. [CrossRef]
3. Scaps, P.; Denis, V. Can organisms associated with live scleractinian corals be used as indicators of coral reef status? *Atoll Res. Bull.* **2008**, *566*, 1–18. [CrossRef]
4. Sorauf, J.E.; Harries, P.J. Rotatory colonies of the corals *Siderastrea radians* and *Solenastraea* ssp. (Cnidaria, Scleractinia), from the Pleistocene Bermont formation, south Florida, USA. *Palaeontology* **2009**, *52*, 111–126. [CrossRef]
5. Kázmér, M.; Taborosi, D. Bioerosion on the small scale–examples from the tropical and subtropical littoral. *Hantkeniana* **2012**, *7*, 37–94.
6. Bagur, M.; Richardson, C.A.; Gutiérrez, J.L.; Arribas, L.P.; Doldan, M.S.; Palomo, M.G. Age, growth and mortality in four populations of the boring bivalve *Lithophaga patagonica* from Argentina. *J. Sea Res.* **2013**, *81*, 49–56. [CrossRef]
7. Glynn, P.W.; Manzello, D.P. Bioerosion and coral reef growth: A dynamic balance. In *Coral Reefs in the Anthropocene*; Birkeland, C., Ed.; Springer: Dordrecht, The Netherlands, 2015; pp. 67–97. [CrossRef]
8. Ricci, S.; Perasso, C.S.; Antonelli, F.; Petriaggi, B.D. Marine bivalves colonizing Roman artefacts recovered in the Gulf of Pozzuoli and in the Blue Grotto in Capri (Naples, Italy): Boring and nestling species. *Int. Biodeter. Biodegrad.* **2015**, *98*, 89–100. [CrossRef]
9. Wizemann, A.; Nandini, S.D.; Stuhldreier, I.; Sánchez-Noguera, C.; Wisshak, M.; Westphal, H.; Rixen, T.; Christian, W.; Reymond, C.E. Rapid bioerosion in a tropical upwelling coral reef. *PLoS ONE* **2018**, *13*, e0202887. [CrossRef]
10. MolluscaBase eds. MolluscaBase. Lithophaginae H. Adams & A. Adams, 1857. World Register of Marine Species. 2022. Available online: https://www.marinespecies.org/aphia.php?p=taxdetails&id=510723 (accessed on 20 April 2022).
11. Zann, L.P. *Living Together in the Sea*; T.F.H. Publications: Neptune, NY, USA, 1980; p. 416.

12. Highsmith, R.C. Coral bioerosion: Damage relative to skeletal density. *Am. Nat.* **1981**, *117*, 193–198. [CrossRef]
13. Hutchings, P.A. Biological destruction of coral reefs. *Coral Reefs* **1986**, *4*, 239–252. [CrossRef]
14. Scott, P.J.B.; Risk, M.J. The effect of *Lithophaga* (Bivalvia: Mytilidae) boreholes on the strength of the coral *Porites lobata*. *Coral Reefs* **1988**, *7*, 145–151. [CrossRef]
15. Rice, M.M.; Maher, R.L.; Correa, A.M.S.; Moeller, H.V.; Lemoine, N.P.; Shantz, A.A.; Burkepile, D.E.; Silbiger, N.J. Macroborer presence on corals increases with nutrient input and promotes parrotfish bioerosion. *Coral Reefs* **2020**, *39*, 409–418. [CrossRef]
16. Wong, K.T.; Tsang, R.H.L.; Ang, P. Did borers make corals more susceptible to a catastrophic disease outbreak in Hong Kong? *Mar. Biodivers.* **2016**, *46*, 325–326. [CrossRef]
17. Yonge, C.M. Adaptation to rock boring in *Botula* and *Lithophaga* (Lamellibranchia, Mytilidae) with a discussion on the evolution of this habit. *J. Cell Sci.* **1955**, *3*, 383–410. [CrossRef]
18. Audino, J.A.; Serb, J.M.; Marian, J.E.A.R. Phylogeny and anatomy of marine mussels (Bivalvia: Mytilidae) reveal convergent evolution of siphon traits. *Zool. J. Linn. Soc.* **2020**, *190*, 592–612. [CrossRef]
19. Soliman, G.N. Ecological aspects of some coral-boring gastropods and bivalves of the northwestern Red Sea. *Am. Zool.* **1969**, *9*, 887–894. [CrossRef]
20. Simone, L.R.L.; Gonçalves, E.P. Anatomical study on *Myoforceps aristatus*, an invasive boring bivalve in S.E. Brazilian coast (Mytilidae). *Pap. Avulsos Zool.* **2006**, *46*, 57–65. [CrossRef]
21. Kleemann, K.H. *Lithophaga* (Bivalvia) from dead coral from the Great Barrier Reef, Australia. *J. Molluscan Stud.* **1984**, *50*, 192–230. [CrossRef]
22. Kleemann, K. Boring and growth in chemically boring bivalves from the Caribbean, Eastern Pacific and Australia's Great Barrier Reef. *Senckenberg. Marit.* **1990**, *22*, 101–154.
23. Gohar, H.A.F.; Soliman, G.N. On three mytilid species boring in living corals. *Publ. Mar. Biol. Sta. Al-Ghardaqa* **1963**, *12*, 65–98.
24. Zottoli, R.A.; Carriker, M.R. Burrow morphology, tube formation, and microarchitecture of shell dissolution by the spionid polychaete *Polydora websteri*. *Mar. Biol.* **1974**, *27*, 307–316. [CrossRef]
25. Hoeksema, B.W. Excavation patterns and spiculae dimensions of the boring sponge *Cliona celata* from the SW Netherlands. *Senckenb. Marit.* **1983**, *15*, 55–85.
26. Liu, P.J.; Hsieh, H.L. Burrow architecture of the spionid polychaete *Polydora villosa* in the corals *Montipora* and *Porites*. *Zool. Stud.* **2000**, *39*, 47–54.
27. Buschbaum, C.; Buschbaum, G.; Schrey, I.; Thieltges, D.W. Shell-boring polychaetes affect gastropod shell strength and crab predation. *Mar. Ecol. Prog. Ser.* **2007**, *329*, 123–130. [CrossRef]
28. Pulido Mantas, T.; Pola, L.; Cerrano, C.; Gambi, M.C.; Calcinai, B. Bioerosion features of boring polydorid polychaetes in the North Adriatic Sea. *Hydrobiologia* **2022**, *849*, 1969–1980. [CrossRef]
29. de Bakker, D.M.; Webb, A.E.; van den Bogaart, L.A.; van Heuven, S.M.A.C.; Meesters, E.H.; van Duyl, F.C. Quantification of chemical and mechanical bioerosion rates of six Caribbean excavating sponge species found on the coral reefs of Curaçao. *PLoS ONE* **2018**, *13*, e0197824. [CrossRef]
30. Stearley, R.F.; Ekdale, A.A. Modern marine bioerosion by macroinvertebrates, northern Gulf of California. *Palaios* **1989**, *4*, 453–467. [CrossRef]
31. Highsmith, R.C. Burrowing by the bivalve mollusc *Lithophaga curta* in the living reef coral *Montipora berryi* and a hypothesis of reciprocal larval recruitment. *Mar. Biol.* **1980**, *56*, 155–162. [CrossRef]
32. Kleemann, K.H. Boring bivalves and their host corals from the Great Barrier Reef. *J. Molluscan Stud.* **1980**, *46*, 13–54. [CrossRef]
33. Nielsen, C. Notes on boring bivalves from Phuket, Thailand. *Ophelia* **1976**, *15*, 141–148. [CrossRef]
34. Cantera, J.R.; Contreras, R. Bivalvos perforadores de esqueletos de corales escleractiniarios en la Isla de Gorgona, Pacífico Colombiano. *Rev. Biol. Trop.* **1988**, *36*, 151–158.
35. Nogueira, M.M.; Menezes, N.M.; Johnsson, R.; Neves, E. The adverse effects of cryptochirid crabs (Decapoda: Brachyura) on *Siderastrea stellata* Verril, 1868 (Anthozoa: Scleractinia): Causes and consequences of cavity establishment. *Cah. Biol. Mar.* **2014**, *55*, 155–162. [CrossRef]
36. Hoeksema, B.W.; van der Meij, S.E.T. Gall crab city: An aggregation of endosymbiotic crabs inhabiting a colossal colony of *Pavona clavus*. *Coral Reefs* **2013**, *32*, 59. [CrossRef]
37. García-Hernández, J.E.; de Gier, W.; van Moorsel, G.W.N.M.; Hoeksema, B.W. The scleractinian *Agaricia undata* as a new host for the coral-gall crab *Opecarcinus hypostegus* at Bonaire, southern Caribbean. *Symbiosis* **2020**, *81*, 303–311. [CrossRef]
38. Lymperaki, M.M.; Hill, C.E.; Hoeksema, B.W. The effects of wave exposure and host cover on coral-associated fauna of a centuries-old artificial reef in the Caribbean. *Ecol. Eng.* **2022**, *176*, 106536. [CrossRef]
39. Hoeksema, B.W.; Harper, C.E.; Langdon-Down, S.J.; van der Schoot, R.J.; Smith-Moorhouse, A.; Spaargaren, R.; Timmerman, R.F. Host range of the coral-associated worm snail *Petaloconchus* sp. (Gastropoda: Vermetidae), a newly discovered cryptogenic pest species in the southern Caribbean. *Diversity* **2022**, *14*, 196. [CrossRef]
40. Hoeksema, B.W.; Timmerman, R.F.; Spaargaren, R.; Smith-Moorhouse, A.; van der Schoot, R.J.; Langdon-Down, S.J.; Harper, C. Morphological modifications and injuries of corals caused by feather duster worms (Sabellidae: *Anamobaea* sp.) in the Caribbean. *Diversity* **2022**, *14*, 332. [CrossRef]
41. Fang, L.-S.; Shen, P. A living mechanical file: The burrowing mechanism of the coral-boring bivalve *Lithophaga nigra*. *Mar. Biol.* **1988**, *97*, 349–354. [CrossRef]

42. Yahel, G.; Marie, D.; Beninger, P.G.; Eckstein, S.; Genin, A. In situ evidence for pre-capture qualitative selection in the tropical bivalve *Lithophaga simplex*. *Aquat. Biol.* **2009**, *6*, 235–246. [CrossRef]

43. Chan, B.K.K.; Tan, J.C.H.; Ganmanee, M. Living in a growing host: Growth pattern and dwelling formation of the scallop *Pedum spondyloideum* in massive *Porites* spp. corals. *Mar. Biol.* **2020**, *167*, 95. [CrossRef]

44. Scaps, P. Association between the scallop *Pedum spondyloideum* (Bivalvia: Pteriomorphia: Pectinidae) and scleractinian coralsfrom Nosy Be, Madagascar. *Cah. Biol. Mar.* **2020**, *61*, 73–80. [CrossRef]

45. Scott, P.J.B. Aspects of living coral associates in Jamaica. In Proceedings of the 5th International Coral Reef Congress, Tahiti, France, 27 May–1 June 1985; Volume 5, pp. 345–350.

46. Scott, P.J.B. A new species of *Lithophaga* (Bivalvia: Lithophaginae) boring corals in the Caribbean. *J. Molluscan Stud.* **1986**, *52*, 55–61. [CrossRef]

47. Scott, P.J.B. Associations between corals and macro-infaunal invertebrates in Jamaica, with a list of Caribbean and Atlantic coral associates. *Bull. Mar. Sci.* **1987**, *40*, 271–286.

48. Scott, P.J.B. Distribution, habitat and morphology of the Caribbean coral and rock-boring bivalve, *Lithophaga bisulcata* (d'Orbigny) (Mytilidae: Lithophaginae). *J. Molluscan Stud.* **1988**, *5*, 83–95. [CrossRef]

49. Scott, P.J.B. Initial settlement behaviour and survivorship of *Lithophaga bisulcata* (d'Orbigny) (Mytilidae: Lithophaginae). *J. Molluscan Stud.* **1988**, *54*, 97–108. [CrossRef]

50. Vinagre, C.; Silva, R.; Mendonça, V.; Flores, A.A.V.; Baeta, A.; Marques, J.C. Food web organization following the invasion of habitat-modifying *Tubastraea* spp. corals appears to favour the invasive borer bivalve *Leiosolenus aristatus*. *Ecol. Ind.* **2018**, *85*, 1204–1209. [CrossRef]

51. Valentich-Scott, P.; Dinesen, G.E. Rock and coral boring Bivalvia (Mollusca) of the middle Florida Keys, USA. *Malacologia* **2004**, *46*, 339–354.

52. Bromley, R.G. Biocrosion of Bermuda reefs. *Paleogeogr. Paleoclimatol. Paleoecol.* **1978**, *23*, 169–197. [CrossRef]

53. McCloskey, L.R. The dynamics of the community associated with a marine scleractinian coral. *Int. Rev. Gesamt. Hydrobiol.* **1970**, *55*, 13–81. [CrossRef]

54. Oigman-Pszczol, S.S.; Creed, J.C. Distribution and abundance of fauna on living tissues of two Brazilian hermatypic corals (*Mussismilia hispida* (Verril, 1902) and *Siderastrea stellata* Verril, 1868). *Hydrobiologia* **2006**, *563*, 143–154. [CrossRef]

55. Blanchon, P.; Perry, C.T. Taphonomic differentiation of *Acropora palmata* facies in cores from Campeche Bank reefs, Gulf of México. *Sedimentology* **2004**, *51*, 53–76. [CrossRef]

56. Donovan, S.K.; Hensley, C. *Gastrochaenolites* Leymerie in the Cenozoic of the Antillean region (review). *Ichnos* **2006**, *13*, 11–19. [CrossRef]

57. Kleemann, K.H. *Gastrochaenolites hospitium* sp. nov., trace fossil by a coral-associated boring bivalve from the Eocene and Miocene of Austria. *Geol. Carpath.* **2009**, *60*, 339–342. [CrossRef]

58. Wisshak, M.; Knaust, D.; Bertling, M. Bioerosion ichnotaxa: Review and annotated list. *Facies* **2019**, *65*, 24. [CrossRef]

59. Bassi, D.; Braga, J.C.; Owada, M.; Aguirre, J.; Lipps, J.H.; Takayanagi, H.; Iryu, Y. Boring bivalve traces in modern reef and deeper-water macroid and rhodolith beds. *Prog. Earth Planet. Sci.* **2020**, *7*, 41. [CrossRef]

60. Mokady, O.; Bonar, D.B.; Arazi, G.; Loya, Y. Coral host specificity in settlement and metamorphosis of the date mussel *Lithophaga lessepsiana* (Vaillant, 1865). *J. Exp. Mar. Biol. Ecol.* **1991**, *146*, 205–216. [CrossRef]

61. Mokady, O.; Arazi, G.; Bonar, D.B.; Loya, Y. Settlement and metamorphosis specificity of *Lithophaga simplex* Iredale (Bivalvia: Mytilidae) on Red Sea corals. *J. Exp. Mar. Biol. Ecol.* **1992**, *162*, 243–251. [CrossRef]

62. Liu, J.C.W.; Hoeg, J.T.; Chan, B.K.K. How do coral barnacles start their life in their hosts? *Biol. Lett.* **2016**, *12*, 20160124. [CrossRef]

63. Dreyer, N.; Tsai, P.-C.; Olesen, J.; Kolbasov, G.A.; Høeg, J.T.; Chan, B.K.K. Independent and adaptive evolution of phenotypic novelties driven by coral symbiosis in barnacle larvae. *Evolution* **2022**, *76*, 139–157. [CrossRef]

64. Marsden, J.R. Coral preference behaviour by planktotrophic larvae of *Spirobranchus giganteus corniculatus* (Serpulidae: Polychaeta). *Coral Reefs* **1987**, *6*, 71–74. [CrossRef]

65. Marsden, J.R.; Conlin, B.E.; Hunte, W. Habitat selection in the tropical polychaete *Spirobranchus giganteus*. *Mar. Biol.* **1990**, *104*, 93–99. [CrossRef]

66. Moretzsohn, F.; Tsuchyia, M. Preliminary survey of the coral boring Bivalvia fauna of Okinawa, southern Japan. In Proceedings of the 7th International Coral Reef Symposium, Guam, 22–26 June 1992; Volume 1, pp. 404–412.

67. Mokady, O.; Rozenblatt, S.; Graur, D.; Loya, Y. Coral-host specificity of Red Sea *Lithophaga* bivalves: Interspecific and intraspecific variation in 12S mitochondrial. *Mol. Mar. Biol. Biotechnol.* **1994**, *3*, 158–164. [PubMed]

68. Kleemann, K.H. Association of coral and boring bivalves: Lizard Island (Great Barrier Reef, Australia) versus Safaga (N Red Sea). *Beitr. Paläontol.* **1995**, *20*, 31–39.

69. Kleemann, K.; Hoeksema, B.W. *Lithophaga* (Bivalvia: Mytilidae), including a new species, boring in mushroom corals (Scleractinia: Fungiidae) at South Sulawesi, Indonesia. *Basteria* **2002**, *66*, 11–24.

70. Hoeksema, B.W.; Kleemann, K. New records of *Fungiacava eilatensis* Goreau et al., 1968 (Bivalvia: Mytilidae) boring in Indonesian mushroom corals (Scleractinia: Fungiidae). *Basteria* **2002**, *66*, 25–30.

71. Mohammed, T.A.; Yassien, M.H. Bivalve assemblages on living coral species in the Northern Red Sea, Egypt. *J. Shellfish Res.* **2008**, *27*, 1217–1223. [CrossRef]

72. Owada, M. The first record of *Leiosolenus simplex* (Iredale, 1939) (Bivalvia: Mytilidae) Boring into *Plesiastrea versipora* from Minamata Bay in Japan. *Venus* **2008**, *67*, 81–84.

73. Owada, M.; Hoeksema, B.W. Molecular phylogeny and shell microstructure of *Fungiacava eilatensis* Goreau et al. 1968, boring into mushroom corals (Scleractinia: Fungiidae), in relation to other mussels (Bivalvia: Mytilidae). *Contrib. Zool.* **2011**, *80*, 169–178. [CrossRef]

74. Kleemann, K.; Maestrati, P. Pacific *Lithophaga* (Bivalvia, Mytilidae) from recent French expeditions with the description of two new species. *Boll. Malacol.* **2012**, *48*, 73–102.

75. Printrakoon, C.; Yeemin, T.; Valentich-Scott, P. Ecology of endolithic bivalve mollusks from Ko Chang, Thailand. *Zool. Stud.* **2016**, *55*, 50. [CrossRef]

76. Liu, J.; Liu, H.; Zhang, H. Phylogeny and evolutionary radiation of the marine mussels (Bivalvia: Mytilidae) based on mitochondrial and nuclear genes. *Mol. Phylogenet. Evol.* **2018**, *126*, 233–240. [CrossRef] [PubMed]

77. Stella, J.S.; Pratchett, M.S.; Hutchings, P.A.; Jones, G.P. Coral-associated invertebrates: Diversity, ecology importance and vulnerability to disturbance. *Oceanogr. Mar. Biol. Ann. Rev.* **2011**, *49*, 43–104.

78. Hoeksema, B.W.; van der Meij, S.E.T.; Fransen, C.H.J.M. The mushroom coral as a habitat. *J. Mar. Biol. Assoc.* **2012**, *92*, 647–663. [CrossRef]

79. Hoeksema, B.W. The hidden biodiversity of tropical coral reefs. *Biodiversity* **2017**, *18*, 8–12. [CrossRef]

80. Hoeksema, B.W.; van Beusekom, M.; ten Hove, H.A.; Ivanenko, V.N.; van der Meij, S.E.T.; van Moorsel, G.W.N.M. *Helioseris cucullata* as a host coral at St. Eustatius. Dutch Caribbean. *Mar. Biodivers.* **2017**, *47*, 71–78. [CrossRef]

81. Montano, S. The extraordinary importance of coral-associated fauna. *Diversity* **2020**, *12*, 357. [CrossRef]

82. Ng, P.K.L.; Meyer, C. A new species of pea crab of the genus *Serenotheres* Ahyong & Ng, 2005 (Crustacea, Brachyura, Pinnotheridae) from the date mussel *Leiosolenus* Carpenter, 1857 (Mollusca, Bivalvia, Mytilidae, Lithophaginae) from the Solomon Islands. *ZooKeys* **2016**, *623*, 31–41. [CrossRef]

83. de Gier, W.; Becker, C. A Review of the ecomorphology of pinnotherine pea crabs (Brachyura: Pinnotheridae), with an updated list of symbiont-host associations. *Diversity* **2020**, *12*, 431. [CrossRef]

Interesting Images

Feeding Behavior of *Coralliophila* sp. on Corals Affected by Caribbean Ciliate Infection (CCI): A New Possible Vector?

Simone Montano [1,2,*], Greta Aeby [3], Paolo Galli [1,2] and Bert W. Hoeksema [4,5]

[1] Department of Earth and Environmental Sciences (DISAT), University of Milan–Bicocca, Piazza della Scienza, 20126 Milan, Italy; paolo.galli@unimib.it
[2] MaRHE Center (Marine Research and High Education Center), Magoodhoo Island, Faafu Atoll 12030, Maldives
[3] Department of Biological and Environmental Sciences, Qatar University, Doha 2713, Qatar; greta@hawaii.edu
[4] Taxonomy, Systematics and Geodiversity Group, Naturalis Biodiversity Center, 2300 RA Leiden, The Netherlands; bert.hoeksema@naturalis.nl
[5] Groningen Institute for Evolutionary Life Sciences, University of Groningen, 9700 CC Groningen, The Netherlands
* Correspondence: simone.montano@unimib.it

Abstract: Coral reefs in the Caribbean are known to be affected by many coral diseases, yet the ecology and etiology of most diseases remain understudied. The Caribbean ciliate infection (CCI) caused by ciliates belonging to the genus *Halofolliculina* is a common disease on Caribbean reefs, with direct contact considered the most likely way through which the ciliates can be transmitted between infected and healthy colonies. Here we report an observation regarding a *Coralliophila* sp. snail feeding in proximity to a cluster of ciliates forming the typical disease band of CCI. The result of this observation is twofold. The feeding behavior of the snail may allow the passive attachment of ciliates on the body or shell of the snail resulting in indirect transport of the ciliates among colonies, which makes it eligible as a possible disease vector. Alternatively, the lesions created from snail feeding may enhance the progression of the ciliates already present on the coral as well as promoting additional infections allowing pathogens to enter through the feeding scar.

Keywords: coral disease; *Halofolliculina*; transmission mechanism; Bonaire; *Acropora*

Citation: Montano, S.; Aeby, G.; Galli, P.; Hoeksema, B.W. Feeding Behavior of *Coralliophila* sp. on Corals Affected by Caribbean Ciliate Infection (CCI): A New Possible Vector? . *Diversity* **2022**, *14*, 363. https://doi.org/10.3390/d14050363

Academic Editors: Andrew Bauman and Harilaos Lessios

Received: 7 April 2022
Accepted: 2 May 2022
Published: 4 May 2022

Publisher's Note: MDPI stays neutral with regard to jurisdictional claims in published maps and institutional affiliations.

Coral diseases represent a serious threat for coral reefs worldwide, with the Caribbean considered a "hotspot" of disease outbreaks [1]. Currently, the coral reefs of the Caribbean are experiencing an outbreak of stony coral tissue loss disease (SCTLD) that originated on the reefs of Florida in 2014 [2] and is spreading across the Caribbean resulting in extensive colony mortality [3–6]. Despite the devastating effect this disease has had on coral reefs, our understanding of its ecology, pathogenesis, and etiologies is limited, making it difficult for resource managers to make decisions on how best to maintain these critical resources. As example, understanding disease spread among individual colonies or coral populations would be a critical factor in developing management actions to slow down or stop disease and the resultant mortality, yet disease transmission dynamics are still understudied [7].

Many studies have suggested that coral-feeding animals can promote disease transmission among colonies on a reef. Several families of coral reef fishes have been observed in the field feeding on coral disease lesions including black band disease (BBD) [8,9], brown band disease (BrB) [10] and stony coral tissue loss disease [11]. Corallivores feeding on disease lesions could transmit the pathogen by subsequently feeding or defaecating on a non-diseased colony and this has been suggested as a mechanism of black band disease transfer by butterflyfishes [9,12] as well as spreading the trematode parasite that causes *Porites trematodiasis* in Hawaii [13]. Numerous types of invertebrate corallivores (snails, nudibranchs, fireworm, crown-of-thorns seastars) have also been implicated in disease transmission, either directly [1,14,15] or indirectly via feeding scars which subsequently develop disease [16–21].

Halofolliculinid ciliates can cause progressive tissue loss on corals; this disease is termed skeletal eroding band in the Indo-Pacific and Caribbean ciliate infection in the Caribbean [22]. Caribbean ciliate infection (CCI) was first reported in 2006 [23], and it can affect ~4 to 8 % of corals as observed in Venezuela and Curaçao [22]. It manifests as a dark-grey band 1–10 cm thick, located at the interface between recently exposed skeleton and apparently healthy coral tissue showing the characteristic spotted appearance of the clustering ciliates [24,25] (Figure 1a,b). *Halofolliculina* ciliates have a life cycle represented by two distinct phases: a sessile ciliate (encased within a lorica), and a motile larval phase. During replication, the de-differentiation of the sessile feeding trophont results into a simple motile phase, which then divides asexually into two motile swarmers that may move using ciliary locomotion and disperse [26]. Transmission of ciliate infection among coral colonies occurs on direct contact between a healthy and infected colony [26] and through the water column if the health colony has a prior injury of any origin [27,28]. Ciliates at the sessile stage have also been found embedded in the shells of a number of corallivorous gastropods which may serve as passive vectors of the disease [22].

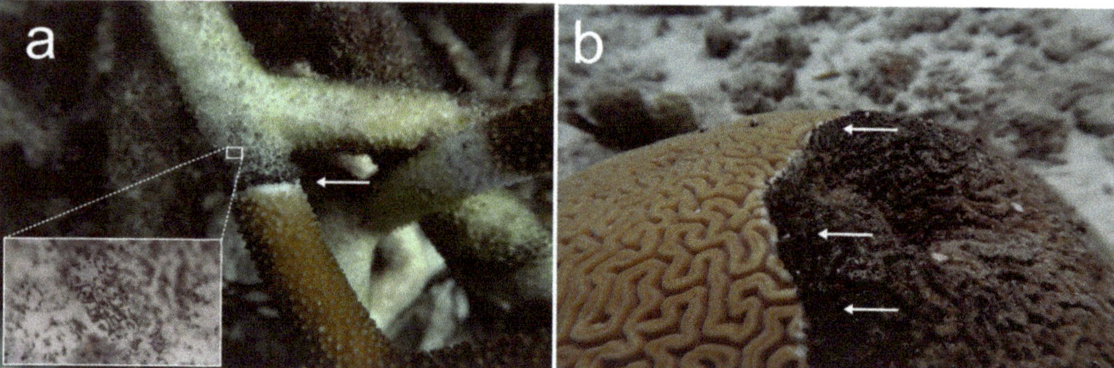

Figure 1. It shows the typical appearance of corals affected by CCI. The arrows indicate the cluster of ciliates forming CCI dark-grey bands located at the interface between recently exposed skeleton and apparently healthy coral tissue on (**a**) *Acropora cervicornis* and (**b**) *Diploria labyrinthiformis*.

Here we report on an observation made in Bonaire in 2019, in which a *Coralliophila* sp. snail was observed feeding on coral tissue at the edge of the cluster of ciliates forming typical disease bands of CCI (Figure 2a,b). The snail species in question is probably *C. galea* (Dillwyn, 1823), previously misidentified as *C. abbreviata* (Lamarck, 1816), which is so far the only *Coralliophila* species reported from Caribbean *Acropora* spp. [29–31]. This observation creates the possibility that transmission of CCI may also be facilitated by snail activities. Lesions created by snail predation may open-up wounds in the coral which can then be colonized by the ciliates at the swarmer stage. Alternatively, snails are attracted to injured coral tissue [32] so coral lesions created by CCI could attract snails and allow passive attachment of ciliates on the body, or shell, of the snail resulting in indirect transport of the ciliates among colonies. *Coralliophila* species have been implicated as a potential vector of white band disease [15,33], white pox disease [1], and white plague disease [34] in the Caribbean, as well as to disease development in *Porites cylindrica* in the Indo-Pacific [20]. Our observation adds to the growing body of evidence on the role that snails play in disease transmission, however, the extent to which *Coralliophila* may be involved in the pathogenesis of halofolliculinid ciliate infection in Bonaire needs further investigation.

Figure 2. (**a**) *Coralliophila* sp. snail feeding in proximity of cluster of ciliates forming the typical disease band of CCI; (**b**) close-up of the snail's feeding behavior.

Author Contributions: Investigation, S.M., B.W.H.; data curation, G.A.; writing—original draft preparation, S.M., G.A., B.W.H.; supervision, P.G. All authors have read and agreed to the published version of the manuscript.

Funding: Fieldwork at Bonaire was supported by the World Wildlife Fund (WWF) Netherlands Biodiversity Fund, the Treub Maatschappij—Society for the Advancement of Research in the Tropics, and by the Nature of the Netherlands program of Naturalis Biodiversity Center.

Acknowledgments: S.M. is grateful to the Naturalis Biodiversity Center for providing Martin Fellowships, which supported fieldwork in Bonaire (2019). We are grateful to the Stichting Nationale Parken (STINAPA) and Dutch Caribbean Nature Alliance (DCNA) at Bonaire for assistance in the submission of the research proposal and the research permit. A special thanks to the Magnificent 7 team for its unforgettable support.

Conflicts of Interest: The authors declare no conflict of interest.

References

1. Sutherland:, K.P.; Shaban, S.; Joyner, J.L.; Porter, J.W.; Lipp, E.K. Human pathogen shown to cause disease in the threatened eklhorn coral *Acropora palmata*. *PLoS ONE* **2011**, *6*, e23468. [CrossRef] [PubMed]

2. Precht, W.F.; Gintert, B.E.; Robbart, M.L.; Fura, R.; Van Woesik, R. Unprecedented disease-related coral mortality in Southeastern Florida. *Sci. Rep.* **2016**, *6*, 31374. [CrossRef] [PubMed]

3. Alvarez-Filip, L.; Estrada-Saldívar, N.; Pérez-Cervantes, E.; Molina-Hernandez, A.; Gonzalez-Barrios, F.J. A rapid spread of the stony coral tissue loss disease outbreak in the Mexican Caribbean. *PeerJ* **2019**, *7*, e8069. [CrossRef] [PubMed]

4. Meiling, S.; Muller, E.M.; Smith, T.B.; Brandt, M.E. 3D photogrammetry reveals dynamics of stony coral tissue loss disease (SCTLD) lesion progression across a thermal stress event. *Front. Mar. Sci.* **2020**, *7*, 1128. [CrossRef]

5. Heres, M.M.; Farmer, B.H.; Elmer, F.; Hertler, H. Ecological consequences of stony coral tissue loss disease in the Turks and Caicos Islands. *Coral Reefs* **2021**, *40*, 609–624. [CrossRef]

6. Thome, P.E.; Rivera-Ortega, J.; Rodríguez-Villalobos, J.C.; Cerqueda-García, D.; Guzmán-Urieta, E.O.; García-Maldonado, J.Q.; Carabantes, N.; Jordán-Dahlgren, E. Local dynamics of a white syndrome outbreak and changes in the microbial community associated with colonies of the scleractinian brain coral *Pseudodiploria strigosa*. *PeerJ* **2021**, *9*, e10695. [CrossRef]

7. Shore, A.; Caldwell, J.M. Modes of coral disease transmission: How do diseases spread between individuals and among populations? *Mar. Biol.* **2019**, *166*, 1–14. [CrossRef]

8. Cole, A.J.; Chong-Seng, K.M.; Pratchett, M.S.; Jones, G.P. Coral-feeding fishes slow progression of black-band disease. *Coral Reefs* **2009**, *28*, 965. [CrossRef]

9. Chong-Seng, K.M.; Cole, A.J.; Pratchett, M.S.; Willis, B.L. Selective feeding by coral reef fishes on coral lesions associated with brown band and black band disease. *Coral Reefs* **2011**, *30*, 473–481. [CrossRef]

10. Nicolet, K.J.; Hoogenboom, M.O. The corallivorous invertebrate *Drupella* aids in transmission of brown band disease on the Great Barrier Reef. *Coral Reefs* **2013**, *32*, 585–595. [CrossRef]

11. Noonan, K.R.; Childress, M.J. Association of butterflyfishes and stony coral tissue loss disease in the Florida Keys. *Coral Reefs* **2020**, *39*, 1581–1590. [CrossRef]

12. Aeby, G.S.; Santavy, D.L. Factors affecting susceptibility of the coral *Montastraea faveolata* to black-band disease. *Mar. Ecol. Prog. Ser.* **2006**, *318*, 103–110. [CrossRef]

13. Aeby, G.S. Interactions of the Digenetic Trematode, *Podocotyloides stenometra*, with Its Coral Intermediate Host and Butterflyfish Definitive Host: Ecology and Evolutionary Implications. .Ph.D. Thesis, University of Hawaii at Manoa, Honolulu, HI, USA, 1998.

14. Sussman, M.; Loya, Y.; Fine, M.; Rosenberg, E. The marine fireworm *Hermodice carunculata* is a winter reservoir and spring-summer vector for the coral-bleaching pathogen *Vibrio shiloi*. *Environ. Microbiol.* **2003**, *5*, 250–255. [CrossRef] [PubMed]

15. Williams, D.E.; Miller, M.W. Coral disease outbreak: Pattern, prevalence and transmission in *Acropora cervicornis*. *Mar. Ecol. Prog. Ser.* **2005**, *301*, 119–128. [CrossRef]

16. Dalton, S.J.; Godwin, S. Progressive coral tissue mortality following predation by a corallivorous nudibranch (*Phestilla* sp). *Coral Reefs* **2006**, *25*, 529. [CrossRef]

17. Nugues, M.M.; Bak, R.P.M. Brown-band syndrome on feeding scars of the crown-of-thorn starfish *Acanthaster planci*. *Coral Reefs* **2009**, *28*, 507–510. [CrossRef]

18. Rypien, K.L.; Baker, D.M. Isotopic labeling and antifungal resistance as tracers of gut passage of the sea fan pathogen *Aspergillus sydowii*. *Dis. Aquat. Organ.* **2009**, *86*, 1–7. [CrossRef]

19. Katz, S.M.; Pollock, F.J.; Bourne, D.G.; Willis, B.L. Crown-of-thorns starfish predation and physical injuries promote brown band disease on corals. *Coral Reefs* **2014**, *33*, 705–716. [CrossRef]

20. Raymundo, L.J.; Work, T.M.; Miller, R.L.; Lozada-Misa, P.L. Effects of *Coralliophila violacea* on tissue loss in the scleractinian corals *Porites* spp. depend on host response. *Dis. Aquat. Organ.* **2016**, *119*, 75–83. [CrossRef]

21. Nicolet, K.J.; Chong-Seng, K.M.; Pratchett, M.S.; Willis, B.L.; Hoogenboom, M.O. Predation scars may influence host susceptibility to pathogens: Evaluating the role of corallivores as vectors of coral disease. *Sci. Rep.* **2018**, *8*, 5258. [CrossRef]

22. Page, C.A.; Cróquer, A.; Bastidas, C.; Rodríguez, S.; Neale, S.J.; Weil, E.; Willis, B.L. Halofolliculina ciliate infections on corals (skeletal eroding disease). In *Diseases of Coral*; Woodley, C.M., Downs, C.A., Bruckner, A.W., Porter, J.W., Galloway, S.B., Eds.; Wiley-Blackwell: Hoboken, NJ, USA, 2016; pp. 361–375.

23. Cróquer, A.; Bastidas, C.; Lipscomb, D. Folliculinid ciliates: A new threat to Caribbean corals? *Dis. Aquat. Org.* **2006**, *69*, 75–78. [CrossRef] [PubMed]

24. Cróquer, A.; Bastidas, C.; Lipscomp, D.; Rodríguez-Martínez, R.E.; JordanDahlgren, E.; Guzman, H.M. First report of folliculinid ciliates affecting Caribbean scleractinian corals. *Coral Reefs* **2006**, *25*, 187–191. [CrossRef]

25. Montano, S.; Maggioni, D.; Liguori, G.; Arrigoni, R.; Berumen, M.L.; Seveso, D.; Galli, P.; Hoeksema, B.W. Morpho-molecular traits of Indo-Pacific and Caribbean *Halofolliculina* ciliate infections. *Coral Reefs* **2020**, *39*, 375–386. [CrossRef]

26. Antonius, A.A.; Lipscomb, D. First protozoan coral-killer identified in the Indo-Pacific. *Atoll Res. Bull.* **2001**, *481*, 1–21. [CrossRef]

27. Page, C.A.; Willis, B.L. Epidemiology of skeletal eroding band on the Great Barrier Reef and the role of injury in the initiation of this widespread coral disease. *Coral Reefs* **2008**, *27*, 257–272. [CrossRef]

28. Rodríguez, S.; Croquer, A.; Guzmàn, H.M.; Bastidas, C. A mechanism of transmission and factors affecting coral susceptibility to *Halofolliculina* sp. infection. *Coral Reefs* **2009**, *28*, 67–77. [CrossRef]

29. Knowlton, N.; Lang, J.C.; Keller, B.D. Fates of staghorn coral isolates on hurricane-damaged reefs in Jamaica: The role of predators. In Proceedings of the 6th International Coral Reef Symposium, Townsville, Australia, 8–12 August 1988; Volume 2, pp. 83–88.

30. Hayes, J.A. Distribution, movement and impact of the corallivorous gastropod *Coralliophila abbreviata* (Lamarck) on a Panamánian patch reef. *J. Exp. Mar. Biol. Ecol.* **1990**, *142*, 25–42. [CrossRef]

31. Potkamp, G.; Vermeij, M.J.A.; Hoeksema, B.W. Genetic and morphological variation in corallivorous snails (*Coralliophila* spp.) living on different host corals at Curaçao, southern Caribbean. *Contrib. Zool.* **2017**, *86*, 111–144. [CrossRef]

32. Morton, B.; Blackmore, G.; Kwok, C.T. Corallivory and prey choice by *Drupella rugosa* (Gastropoda: Muricidae) in Hong Kong. *J. Molluscan Stud.* **2002**, *68*, 217–223. [CrossRef]

33. Gignoux-Wolfsohn, S.A.; Marks, C.J.; Vollmer, S.V. White band disease transmission in the threatened coral *Acropora cervicornis*. *Sci. Rep.* **2012**, *2*, 804. [CrossRef]

34. Clemens, E.; Brandt, M.E. Multiple mechanisms of transmission of the Caribbean coral disease white plague. *Coral Reefs* **2015**, *34*, 1179–1188. [CrossRef]

Interesting Images

First Report of Potential Coral Disease in the Coral Hatchery of Thailand

Suppakarn Jandang [1], **Dewi E. Bulan** [1,2], **Suchana Chavanich** [1,3,*], **Voranop Viyakarn** [1,3], **Kornrawee Aiemsomboon** [1] and **Naraporn Somboonna** [4]

1. Reef Biology Research Group, Department of Marine Science, Faculty of Science, Chulalongkorn University, Bangkok 10330, Thailand; bio_aqa@hotmail.com (S.J.); dewi.embong@fpik.unmul.ac.id (D.E.B.); voranop.v@chula.ac.th (V.V.); kornrawee.a@gmail.com (K.A.)
2. Department of Aquatic Resources Management, Faculty of Fisheries and Marine Science, University of Mulawarman, Samarinda 75119, Indonesia
3. Aquatic Resources Research Institute, Chulalongkorn University, Bangkok 10330, Thailand
4. Department of Microbiology, Faculty of Science, Chulalongkorn University, Bangkok 10330, Thailand; Naraporn.S@chula.ac.th
* Correspondence: suchana.c@chula.ac.th; Tel.: +662-218-5394; Fax: +662-255-0780

Abstract: In this study, coral disease was first reported in the coral hatchery in Thailand. Disease were usually found on corals aged two to five years old during the months of November to December of each year. To identify bacterial strains, culture-based methods for strain isolation and molecular techniques of the 16S rRNA gene analysis were used. The resuts showed that the dominant genera of bacteria in diseased corals were *Vibrio* spp. (comprising 41.01% of the isolates). The occurrence of the disease in the coral hatchery can have a significant effect on the health and survival of juvenile corals before being transplanted to natural reefs for restoration.

Keywords: coral; culture; disease; Thailand; temperature

Citation: Jandang, S.; Bulan, D.E.; Chavanich, S.; Viyakarn, V.; Aiemsomboon, K.; Somboonna, N. First Report of Potential Coral Disease in the Coral Hatchery of Thailand. *Diversity* **2022**, *14*, 18. https://doi.org/10.3390/d14010018

Academic Editors: Michael Wink and Simone Montano

Received: 21 August 2021
Accepted: 27 December 2021
Published: 29 December 2021

Publisher's Note: MDPI stays neutral with regard to jurisdictional claims in published maps and institutional affiliations.

Coral restoration has long been implemented in Thailand. However, most programs use asexual reproduction methods to produce new corals. In 2008, the first coral hatchery was established at Samea San Island, upper Gulf of Thailand to culture several coral species using a sexual reproduction technique, and to raise corals to an age of five years old before being transplanted to natural reefs. In this study, we report for the first time on the incidence of coral disease found in the hatchery. During the months of November and December, annually since 2015, coral disease has been found on cultured *Platygyra* corals. Diseases were usually found on corals aged two to five years old (Figure 1). More than 100 coral colonies infected by disease (approximately 25% of total corals in the hatchery) were recorded each year.

To identify bacterial strains, culture-based methods for strain isolation and molecular techniques of the 16S rRNA gene analysis were used. Partial sequences of the 16S rRNA gene revealed that the dominant genera of bacteria in diseased corals were *Vibrio* spp. (comprising 41.01% of the isolates, followed by *Bacillus megaterium* 25.28%, *Pseudoalteromonas* spp. 21.35%, *Promicromonospora citrea* 6.74%, and unidentified bacterium 2D804 5.62%). In comparison, healthy corals possessed a small quantity of *Vibrio* spp. (7.76%). These findings indicate that certain bacteria were able to become dominant in coral hosts (e.g., *Vibrio*) while others were drastically reduced or lost (e.g., *Alteromonas* and *Nocardiopsis*) during the low water temperatures when disease was most prevalent. In addition, analysis of the culture-independent bacterial ribosomal intergenic spacer showed the differences in bacterial communities between diseased and healthy corals, which was similar to the findings of Bourne et al. of which *Vibrio* was dominant in the diseased community [1]. The occurrence of dominant *Vibrio* spp. suggested that these bacteria species may be opportunistic pathogens on healthy corals during winter seasons when coral immunity may be reduced due to lower water temperatures [2,3]. In addition to corals, the winter disease scenarios were also found in shrimp and crab species [2,4,5]. In Thailand, coral disease can be found throughout both the Gulf of Thailand and the Andaman Sea, particularly after bleaching events [6,7].

Figure 1. Potential coral disease in the coral hatchery (**a,b**). The red arrows indicate where the coral disease was found.

Our findings are the first to demonstrate the dominance of *Vibrio* and the changing bacterial assemblages in diseased corals in the hatchery during winter or low temperature seasons. The occurrence of the disease in the coral hatchery can have a significant effect on the health and survival of juvenile corals before being transplanted to natural reefs for restoration.

Author Contributions: Conceptualization, S.J., S.C., V.V. and N.S.; identification, D.E.B., S.C. and N.S.; validation, S.C. and N.S.; investigation, S.J., D.E.B., S.C., V.V., K.A. and N.S.; writing—review and editing, S.J., D.E.B., S.C., V.V., K.A. and N.S.; project administration, S.C.; funding administration, S.C., V.V. and N.S. All authors have read and agreed to the published version of the manuscript.

Funding: The funding was provided by Ph.D. scholarship of Royal Golden Jubilee, NRCT-JSPS Core to Core Program, Thailand Science Research and Innovation Fund Chulalongkorn University (CU_FRB65_dis (3)_091_23_23)), and Mubadala Petroleum (Thailand) Limited.

Institutional Review Board Statement: Not applicable.

Informed Consent Statement: Not applicable.

Data Availability Statement: Not applicable.

Acknowledgments: This work was supported by the Plant Genetic Conservation Project under the Royal Initiative of Her Royal Highness Princess Maha Chakri Sirindhorn and the Naval Special Warfare Command, the Royal Thai Navy.

Conflicts of Interest: The authors declare no conflict of interest.

References

1. Bourne, D.G.; Iida, Y.; Uthicke, S.; Smith-Keune, C. Changes in coral-associated microbial communities during a bleaching event. *ISME J.* **2008**, *2*, 350–363. [CrossRef] [PubMed]
2. Gunalan, B.; Soundarapandian, P.; Dinakaran, G.K. The effect of temperature and pH on WSSV infection in cultured marine shrimp. *Penaeus monodon* (Fabricius). *Middle East J. Sci. Res.* **2010**, *5*, 28–33.
3. Toledo-Hernández, C.; Ruiz-Diaz, C.P. The immune responses of the coral. *Invertebr. Surviv. J.* **2014**, *11*, 319–328.
4. Gavish, A.; Shapito, O.H.; Kramarsky-Winter, E.; Vardi, A. Microscale tracking of coral-vibrio interactions. *ISME Commun.* **2021**, *1*, 18. [CrossRef]
5. Bulan, D.E.; Wilantho, S.; Tongsima, V.; Viyakarn, V.; Chavanich, S.; Somboonna, N. Microbial and small eukaryotes associated. with reefs in the upper Gulf of Thailand. *Front. Mar. Sci.* **2018**, *5*, 436. [CrossRef]
6. Chavanich, S.; Viyakarn, V.; Loyjiw, T.; Pattaratamrong, P.; Chankong, A. Mass bleaching of soft coral, *Sarcophyton* spp. in Thailand and the role of temperature and salinity stress. *ICES J. Mar. Sci.* **2009**, *66*, 1515–1519. [CrossRef]
7. Roder, C.; Arif, C.; Bayer, T.; Aranda, M.; Daniels, C.; Shibl, A.; Chavanich, S.; Voolstra, C.R. Bacterial profiling of white plague disease (WPD) in a comparative coral species framework. *ISME J.* **2014**, *8*, 31–39. [CrossRef] [PubMed]

MDPI

St. Alban-Anlage 66

4052 Basel

Switzerland

Tel. +41 61 683 77 34

Fax +41 61 302 89 18

www.mdpi.com

Diversity Editorial Office

E-mail: diversity@mdpi.com

www.mdpi.com/journal/diversity